FIBRES

Other volumes in
The Chemistry in Modern Industry Series

CHEMICAL ENGINEERING: AN INTRODUCTION
J. R. Potter

PHARMACEUTICALS
L. K. Sharp and J. N. T. Gilbert

PLASTICS AND RUBBERS
E. W. Duck

CHEMICAL INDUSTRY: SOCIAL AND
ECONOMIC ASPECTS
F. R. Bradbury and B. G. Dutton

Series Editors

E. W. Duck
M.Sc.(London), Dr.Sc.Nat.(Aachen), F.R.I.C., F.I.R.I.
D. J. Daniels
B.Sc., Ph.D., F.R.I.C.

FIBRES

C. B. CHAPMAN B.Sc., Ph.D.
Senior Research Scientist,
I.C.I. Fibres Ltd.

LONDON
BUTTERWORTHS

THE BUTTERWORTH GROUP

ENGLAND
Butterworth & Co (Publishers) Ltd
London: 88 Kingsway, WC2B 6AB

AUSTRALIA
Butterworths Pty Ltd
Sydney: 586 Pacific Highway, NSW 2067
Melbourne: 343 Little Collins Street, 3000
Brisbane: 240 Queen Street, 4000

CANADA
Butterworth & Co (Canada) Ltd
Toronto: 14 Curity Avenue, 374

NEW ZEALAND
Butterworths of New Zealand Ltd
Wellington: 26–28 Waring Taylor Street, 1

SOUTH AFRICA
Butterworth & Co (South Africa) (Pty) Ltd
Durban: 152–154 Gale Street

First published 1974

© Butterworth & Co (Publishers) Ltd, 1974

ISBN 0 408 70575 2

Printed in England by Hazell Watson & Viney Ltd.

A/677

Preface

The everyday contact with a variety of textile articles and fabrics, and the major importance of the textile and clothing industries, more than justify some elementary consideration of textile fibres and fabrics as part of a general education. The scope of the subject, bridging science, technology, craft, and art, is too wide to be covered in any detail within a single volume. This book is aimed at the beginnings, the production of fibres, the raw, but extremely sophisticated, materials for the textile industry. It is written for the Sixth Form science student who is studying chemistry and who wishes to appreciate something of the application of the subject to the creation of the functional and beautiful materials which surround him or her. The emphasis is heavily upon chemistry and upon man-made fibres, for the synthetic fibres are products of the chemical industry, but it is hoped that the budding engineer and physicist, taking chemistry as a secondary subject, may find some indications of the applications of their expertise which will set them thinking.

It is the hope of the author that the book will provide some general introduction to the basic scientific aspects of the production of fibres, and to the physical and chemical principles underlying the properties of fibres. The chemical synthesis of a potential synthetic fibre is not particularly difficult, and follows from quite simple principles. The synthesis of a radically new and commercially successful fibre is something quite different of course!

The coverage of individual fibres is not complete, and the treatment neither detailed nor mathematical. It is the intention to stimulate as well as to inform, to show the limits of our understanding and abilities, and to indicate the subtleties which can arise in industrial processes which may seem established and simple. After steady evolution over thousands of years, the processing of natural fibres became industrialised, and then within this century the textile industry has been revolutionised by the man-made fibres. Perhaps the exciting days of fibre discovery and invention are past, but the latest developments are perhaps equally revolutionary probings into the traditional areas of fabric making. The future promises to be as exciting.

The author wishes to thank the Directors of I.C.I. Fibres Ltd. for permission to publish, and is indebted to several colleagues who have read the manuscript, and have provided photographs and diagrams. Any opinions expressed are entirely the responsibility of the author. My grateful thanks are also due to the Series Editor, Dr. D. J. Daniels of the University of Bath, Department of Education, for his most patient assistance and constructive criticism, and to Dr. B. C. Shurlock of Butterworths.

Contents

1	Fibres	1
2	The Basic Science of Synthetic Fibres	6
3	Spinning Man-Made Fibres	30
4	The Production and Preparation of Natural Textile Fibres	49
5	Cellulosic Fibres	59
6	Polyamide Fibres	68
7	Polyester Fibres	85
8	Acrylic Fibres	94
9	Polyalkene Fibres	102
10	Elastomeric Fibres	108
11	Inorganic Fibres and Carbon Fibres	113
12	Yarns and Fabrics	118
13	The Future	128
	Index	133

Fibres

This is an unlikely book to read in the bath, so the reader is most probably comfortably enclosed in several assemblies of fibres, sitting on another, and with feet possibly on yet another. Each of these fibrous structures—garment, upholstery, carpet—is so common-place that most of us take them for granted, but each is the end product of a long chain of operations. We choose them with care, we wear them perhaps with pride; we rarely think of what and how they are made! Why make tights of nylon? (What is that anyway?) Why does the crease stay in those trousers? The science, technology, and art of fibres and fabrics present a rich, complex, and intriguing field for work and study—whether it is for a living or for enjoyment.

The history of textile fibres and fabrics is as old as the history of civilisations, so central to our lives are textiles for clothing, bedding and furnishing. The design of fabrics and clothes becomes an art form of a civilisation. The natural fibres, chief amongst them wool, silk, cotton, and linen, have been known and used with skill for thousands of years.

In contrast, the history of man-made fibres is practically a story of the past 100 years, but it is also a fascinating story, especially to the scientist. Man-made fibres have contributed greatly to our present day society, not only to the kind of clothes we wear, but also to our wealth and comfort of living, in the broadest sense of these words. Only the briefest and most inadequate mention of the scientific history is given as an introduction to the following chapter, which is concerned with the fundamentals of fibres.

We cannot define the 'perfect fibre', because so many different things are demanded of a fibre in a very large number of different circumstances of use. A variety of fibres is necessary to satisfy all the many uses, and the tastes and preferences of individuals. But the major fibres which we shall discuss can each satisfy the basic requirements of an incredible number of products.

A fibre can be defined generally as a form of matter of small cross-sectional area and with length greatly exceeding the width. If the length is very long, so that in practice we may regard it as infinite, it is referred to as a continuous filament. A short length, no more than ten centimetres or so, is called staple fibre. The man-made fibres are produced as continuous filaments, which may then be cut up to make staple fibre. Wool and cotton are staple fibres, whilst silk is a natural continuous filament.

The wool fibre has a fine, wavy form, called crimp, which gives the fibre softness and spring-like elasticity. This desirable kind of structure can be given to the man-made straight continuous filaments, and they are then called 'bulked' or 'textured' (Chapter 12).

If a number of filaments are combined together so that they do not separate easily, but have coherence, we have a continuous filament yarn. (A single continuous filament may also be used as a yarn—monofilament—for example for fine, sheer ladies' tights.) Similarly, by twisting the short staple fibres together a yarn is made. A staple yarn is usually softer and warmer than a corresponding straight continuous filament yarn, for the short fibres do not all lie parallel to the length of the yarn, but protrude from the surface, and trap more air between them. Long cylindrical filaments pack more tightly, trap little air, and feel harsh and cold in comparison. Textured continuous filament yarns are more like staple yarns in this respect.

The process of producing yarns from staple is called spinning: the same word is used to describe the methods of manufacturing continuous filament man-made fibres. The context in the book should make it clear which spinning process is implied.

Yarns are used to make fabrics. In this book we are mainly concerned with the making of continuous filaments, and shall only briefly touch upon their conversion to yarn and to fabric. It will be appreciated that a virtually infinite number of yarns can be made by combining together different numbers of fibres or filaments, of different diameters, and of different physical properties. Each yarn has different properties and characteristics, and each can be made up into fabrics in many different combinations and constructions. Hence the infinitely rich variety of textile fabrics!

Before the fabric reaches the customer it is subjected to important 'finishing' treatments which develop and improve its properties and appearance, and improve its dimensional stability to washing and to use. It may be bleached, dyed, or printed. The chemist makes a considerable contribution to this end of the textile business.

A garment or domestic fabric is commonly judged by its appearance (including colour and design) and by touch or 'handle'—how soft or harsh, how warm or cool, how dry or waxy, how resilient it is. A trained textile technologist has a much keener appreciation of these qualities than the average domestic customer, and he will carefully select his yarns and fabric construction to produce something, within a price consideration, which is both aesthetic and functional.

The rapidly growing world population is making ever increasing demands on all resources, including fibres. The world production of raw cotton was about 3×10^9 kg at the beginning of the century, in 1971 it was about 12×10^9 kg. In the same period the production of wool has doubled from about 8×10^8 to 16×10^8 kg. There was negligible man-made fibre production in 1900. In 1971 about 10×10^9 kg of man-made fibres were produced—about 40% of the total of all fibres. The population of the world has increased from about 1.5×10^9 to about 3.7×10^9 within this time.

These figures illustrate dramatically how the consumption of fibres increases with the rising standard of living, and not in simple direct proportion to numbers. Most of the fibres are used in the developed countries, and clothing and fabrics for the home take up about two-thirds of all consumed, with industrial uses (ropes, tyres, etc.) taking about one-fifth.

The growth of the man-made fibres industry has been quite phenomenal. The world production figures for three major fibres, rayon, nylon, and polyester, since their introduction, are shown graphically in *Figure 1.1*. These figures have been 'smoothed' to

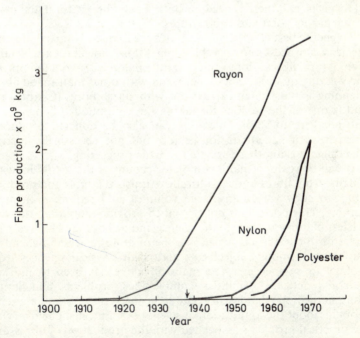

Figure 1.1   Annual world production of three man-made fibres

iron out yearly fluctuations; the decline in rayon in recent years may be real, or could be a temporary check which will go as other countries begin or step-up production.

The necessity of man-made fibres to our present society is obvious from the above statistics. Natural fibres are subject to random yearly variations in quality and in availability, and therefore in price. Expansion of production is possible neither easily nor rapidly, and uses up more land. The important technical and economic advantages of man-made fibres over the natural fibres lie in the control of quality, and in the ability to make fibres of whatever diameter, length, and properties are needed. The expansion of production is relatively easy, and price can be stabilised reasonably well.

Coupled with this controlled adaptability of man-made fibres to the needs of the textile processing industry, is their versatility (they can be used for very many different end products), and their novelty. Novel properties make possible new uses, the improvement of performance of existing products, and new fabrics and fashions to meet the needs of our society. How difficult would be mass holiday air travel with the cumbersome fashions and the fabrics of a century ago! But, man-made fibres by no means replace the natural fibres in all respects, and there is need for both. Each fibre, natural and man-made, has its good and bad features.

To get the best of both worlds, fibres are frequently mixed together in yarns; a process known as blending. Man-made fibres are blended with natural fibres—for example, a little nylon with wool to improve the wearing qualities of a carpet—and with other man-made fibres. Blending is most often carried out with staple fibres (Chapter 12), and is one of the reasons for taking the continuous filaments and cutting them to staple. Blending can also be done to produce a cheaper yarn—adequate for its job and not necessarily inferior. Blending adds another variable to textile processing.

It will be realised that a complete account of the textile industry ranging from fibres to the multitude of finished fabrics and products would occupy very many large volumes, and require even more authors. This book does not set out to provide more than the very briefest of introductions to the science and technology of fibres, with stress on those aspects which may be of more interest to chemists. But this enormous subject cuts across all physical sciences from biology to engineering. The main objective has been to indicate general practice, possibilities, complexities, problems, and limitations, in the hope that it will excite some interest and stimulate some thought about the application of science (and not just chemistry) to the practical problems associated with the production of fibres, and to the better understanding of fibre properties. It has not been the intention to provide answers to questions, but to provoke more questions. Those who would like more detailed and factual knowledge will find many other elementary and advanced books, both of polymer chemistry and of textiles, to help them to a deeper understanding. Much interesting detail of chemical technology is not published, for reasons of industrial security, until it is old, or outdated.

However, science is a practical subject, and much can be learned in a school laboratory about the general properties of fibres and fabrics by quite simple physical tests, and by hand-lens and low power microscopic examination of scraps of different fabrics and garments. See how the fabric is constructed, then how the yarn (or yarns) is assembled, and then examine the fibre, in length and cross-section if possible. The range of chemical work for the school laboratory is perhaps more limited, but the spinning of crude fibres from a melt and from solution (Chapter 3) is possible, and fibres can be identified by solubility and melting behaviour. A combination of physics and chemistry, the effects of exposure to various chemicals and to light on the simple physical properties, for example breaking load, of yarns and fabrics, widens the scope. Some simple qualitative comparisons of dyeing behaviour, and the fastness of dyes to washing, and fading in sunlight can also be made.

A few simple experiments and observations will greatly help the appreciation of this vital, vast, and rich field of activity.

2

The Basic Science of Synthetic Fibres

Since Man first began to make his own clothes by spinning the natural fibres of animal and of vegetable origin into yarns, he must have speculated upon the ability of the spider and silkworm to spin continuous, fine filaments. A written suggestion of the possibility and benefit of making artificial silk fibres was made by Robert Hooke, F.R.S. in 1665 in his book *Micrographia*. The silkworm cocoon can provide up to 800 m of reelable silk, with diameter about 15 µm and a breaking load of perhaps 5×10^3 kgf cm^{-2}. The spider's silk has not been used on any appreciable scale, because it is not as readily accessible, but in fact the spider is the more versatile and produces different silks to perform special functions. She makes a very fine drag line almost as strong as high tenacity nylon but with a high extensibility to take up the momentum of her falling body. The cocoon is wrapped with a weaker but bulked filament to protect the eggs. The sticky spiral of the web is a highly elastic filament.

The synthetic fibre industry has learned how to make fibres with similar properties, but cannot yet match the spider in producing such a range from a common set of amino acids!

Imitation of Nature's fibres was beyond the reach of Man until the scientific and industrial developments of the nineteenth century. Even then chemical knowledge did not permit deliberate synthesis of fibre forming materials, and so early attempts to make fibres concentrated upon conversions of natural products. The first fibres were thus 'regenerated fibres' formed by taking natural cotton or protein, subjecting this to some chemical treatment to bring it into solution, extruding the solution through fine holes, and obtaining fibres from the jets by re-precipitating the original solid in some way.

Soluble cellulose derivatives were produced first; cellulose nitrate (guncotton) was made in 1832 by treating cellulosic materials (cotton, wood, paper) with nitric acid. Audemars, a French scientist, made crude fibres by drawing out a needle from cellulose nitrate solution about 20 years later, but it was 1883 before Joseph Swann extruded the first filaments. Swann was searching for filaments for the new electric lamp, and carbonised them to make carbon fibres, but he also recognised their use as a textile fibre. Chardonnet made cellulose nitrate fibres on a commercial scale in France in 1885. The nitrated material is, of course, highly flammable and usable fibres were obtained by treatment with ammonium sulphide solution to convert the nitrate back to cellulose.

The production of 'artificial silk' as it was called began in France, which had a natural silk industry, but there was probably incomplete appreciation at that time of the really enormous potential of a synthetic fibre industry. However, Chardonnet demonstrated the possibilities, presumably at some profit. Several developments followed, more rapidly now, and the most important discoveries were made by two British chemists, C. F. Cross and E. J. Bevan. They were experts in paper-making and worked in London, as consultants, on the chemistry of cellulose, at that time little understood. They found two methods of converting cellulose cheaply into more attractive soluble derivatives, known as the viscose rayon process (1891) and the cellulose acetate process (1894). A process for spinning fibres from viscose was patented by H. Stearn, and the three worked together to develop the viscose process from the laboratory to a crude pilot plant scale. Both of these processes were eventually brought to an economic industrial scale and gradually ousted all earlier methods. They form the basis for present day production of large quantities of fibres (Chapter 5).

Vegetable fibres are cellulosic in chemical composition, animal fibres are protein. The regeneration of fibres from natural protein on a commercial scale proved more difficult, and it was not until 1935 in Italy that fibres were produced from casein, milk protein. Although several regenerated protein fibres have since been manufactured on a small scale, deriving protein from milk, groundnuts, soyabean, or maize, none has been really successful and no further reference is made to them here. The attraction of protein fibres derives from wool and silk, both of which have highly desirable properties in handling as fabrics, still not fully matched by synthetics. Perhaps some future fibre will be protein-like in its chemical structure, but there is also, of course, the question of the desirability of making fibres from protein which could be used to provide food.

The production of truly synthetic fibres, as opposed to the regenerated ones, began as recently as 1930 or thereabouts.

During the early years of the present century scientists began to acquire the basic knowledge and develop the techniques which enabled them to investigate the molecular structures of the more complex natural products, a labour still very far from complete. Also, the infant plastics and synthetic rubber industries were growing. The work of academic and industrial chemists and physicists on the structure and synthesis of plastics materials, coupled with the investigation of natural rubber, of cellulose, and of wool, silk, and other proteins, brought about the gradual realisation that the molecules of such materials are composed of many thousands of atoms joined together by chemical bonds. Whereas classical organic chemistry is concerned with small, but still complex, molecules, rarely composed of as many as one hundred atoms, Nature presents the challenge that her important constructional materials are composed of truly giant molecules (macromolecules). However, it was recognised that certain basic molecular sub-units existed, containing perhaps ten to twenty atoms, and that the large molecules were built up from many of these units joined end to end. Hence the term 'polymer' (many parts) was coined.

The concept of the macromolecule marks an important step in the development of chemistry. In common with most advances in science, as in all areas of human activity, many people contributed both work and thought over several years, but H. Staudinger was the first to propose clearly synthetic polymeric structures (1920) and to provide logical experimental evidence in support of them. Universal acceptance by chemists was delayed to the 1930s for various historical reasons (scientists in general may be as reluctant to accept new ideas as anyone else) and more cogently because there was little convincing experimental proof.

The fact that natural macromolecules are built up from small basic units pointed the way to laboratory methods of making novel, synthetic polymers, first by making the units—the monomer—and then by coupling them together, both stages being possible by methods known to classical organic chemistry. Outstanding in the realisation and early exploitation of this important principle was W. H. Carothers, who worked for the Du Pont Company (U.S.A.) from 1928 onwards, and produced the synthetic fibre nylon (and other important polymers) which was first marketed in 1938.

Once the principles of polymer structure and synthesis were established, and with the production of the first plastics, synthetic rubbers and synthetic fibres, the foundations were laid by the late 1940s for the rapid rise of giant industries, and for an upsurge of basic and applied research work in the new Polymer Science. This growth is by no means ended: it has been estimated that by the year 2000 the annual production of synthetic fibres and of plastics in the U.K. will be about ten times the level of 1966.

POLYMERISATION REACTIONS

Although many organic reactions are applied in the laboratory to combine monomer units to form polymer molecules, relatively few are employed in the commercial production of synthetic fibres. It is opportune at this point to consider briefly the general classes of polymerisation reaction which are used industrially; further details of the chemistry will be given as individual fibres are described.

Condensation reactions are used to prepare polyamides and polyesters. For example, the reaction of an aliphatic carboxylic acid with a primary amine to form an amide,

$$CH_3CH_2COOH + H_2N(CH_2)_3CH_3$$
$$\rightleftharpoons CH_3CH_2CONH(CH_2)_3CH_3 + H_2O$$

can be used to form a polymer by reacting a dicarboxylic acid with a diamine. Reaction of 1,6-diaminohexane (hexamethylene diamine) with hexane-1,6-dioic acid (adipic acid) produces a polyamide (6.6 nylon),

$$n\text{-}H_2N(CH_2)_6NH_2 + n\text{-}HOOC(CH_2)_4COOH \rightleftharpoons H_2N(CH_2)_6NH$$
$$-[CO(CH_2)_4CONH(CH_2)_6NH]_{n-1}CO(CH_2)_4COOH$$
$$+ (2n-1)H_2O$$

A mixture of different diacids or diamines may be used to produce a polyamide with a randomly varying composition along the polymer molecule. Such a product is described as a copolymer, and co-polymerisation offers an important means of varying the chemical and physical properties of polymeric materials. Instead of using a diamine/diacid mixture certain amino acids may be self-condensed, e.g.

$$2n\text{-}H_2N(CH_2)_5COOH \rightleftharpoons$$
$$H_2N-[(CH_2)_5CONH(CH_2)_5CO]_n-OH + (2n-1)H_2O$$

Polyesters may be prepared by analogous reactions from appropriate dicarboxylic acids and diols, or from hydroxyacids.

The addition reaction of diisocyanates with diols gives poly-urethanes,

$$OCN(CH_2)_6NCO + HO(CH_2)_4OH \longrightarrow$$
$$-CO.NH(CH_2)_6NH.CO.O(CH_2)_4O-$$

This reaction is commonly used as a means of preparing rubbery polymers for spinning to elastic fibres (see Chapter 10).

The addition and condensation reactions given above have one important feature in common: polymerisation proceeds by random reaction of functional groups, resulting in the relatively slow build-up of the relative molecular mass of the polymer throughout the whole

period of the polymerisation process. A relative molecular mass of
15 000–20 000 may be required for fibre spinning, and in order to
achieve this a very high reaction conversion is required—perhaps
as many as 99.5% of the initial functional groups must react. This
requirement imposes severe standards of purity upon the chemicals
used to prepare polymers for fibres by these means; a small proportion of monofunctional impurity will restrict the relative molecular
mass of the polymer by blocking off the ends of some of the molecules
and preventing further growth at these ends. This effect is put to
good use for the control of relative molecular mass by the deliberate
addition of precise quantities of monofunctional reagents.

Ethenyl (or vinyl) monomers, of general structure $CH_2{=}CH.R$,
where R is an organic group or a halogen, may be polymerised by
addition reactions at the double bond. Initiation of polymerisation
is usually by free radicals which are generated by the decomposition
of peroxides or certain azo compounds. A simple initiator system is
the iron(II) ion/hydrogen peroxide redox reaction:

$$Fe^{2+} + H_2O_2 \longrightarrow Fe^{3+} + OH^- + OH^{\cdot}$$

The hydroxyl radical initiates polymerisation, as follows,

$$OH^{\cdot} + CH_2{=}CH.R \longrightarrow HO.CH_2{-}\overset{\cdot}{C}H.R$$

$$HO.CH_2\overset{\cdot}{C}H.R + CH_2{=}CH.R \longrightarrow HO.CH_2{-}\underset{\underset{R}{|}}{CH}{-}CH_2{-}\overset{\cdot}{C}HR$$

The polymer molecule grows by the repeated addition of monomer
to the polymeric free radical. Growth is stopped either by the combination or disproportionation of two radicals, or by hydrogen
abstraction from one of the molecular species present in the reaction
mixture.

Combination:

$$HO(CH_2\underset{\underset{R}{|}}{CH})_n CH_2\overset{\cdot}{C}HR + HO(CH_2\underset{\underset{R}{|}}{CH})_m CH_2\overset{\cdot}{C}HR \longrightarrow$$

$$HO(CH_2\underset{\underset{R}{|}}{CH})_{n+1}(CH_2\underset{\underset{R}{|}}{CH})_{m+1}OH$$

Disproportionation:

$$HO(CH_2\underset{\underset{R}{|}}{CH})_x CH_2\overset{\cdot}{C}HR + HO(CH_2\underset{\underset{R}{|}}{CH})_y CH_2\overset{\cdot}{C}HR \longrightarrow$$

$$HO(CH_2CH)_xCH_2CH_2R + HO(CH_2CH)_yCH{=}CHR$$
$$\quad\quad\ |\qquad\qquad\qquad\qquad\qquad\quad |$$
$$\quad\quad R\qquad\qquad\qquad\qquad\qquad\quad R$$

Hydrogen abstraction:

$$HO(CH_2CH)_zCH_2\overset{\cdot}{C}HR + R'H \longrightarrow HO(CH_2CH)_zCH_2CH_2R + R''$$
$$\quad\ |\qquad\qquad\qquad\qquad\qquad\qquad\qquad |$$
$$\quad R\qquad\qquad\qquad\qquad\qquad\qquad\quad R$$

Control of the relative molecular mass of the polymer molecule can be exercised by the deliberate selection and addition of the reactive species R'H.

Three distinguishing features are of significance in comparison with the non-radical polymerisations described previously.

(1) Once initiated, each polymer molecule builds up rapidly until terminated: complete polymer molecules and monomer exist together.

(2) Very much higher relative molecular masses are generally attained—hundreds of thousands rather than a few tens of thousands.

(3) The polymer molecules are not of high structural regularity because radical addition can occur at either of the two carbon atoms of the double bond in the original monomer:

$$P{-}CH_2\overset{\cdot}{C}HR + CH_2{=}CHR \longrightarrow P{-}CH_2CH{-}CH_2\overset{\cdot}{C}HR$$
$$\qquad\qquad\qquad\qquad\qquad\qquad\qquad\qquad |$$
$$\qquad\qquad\qquad\qquad\qquad\qquad\qquad\quad R$$

$$\text{or} \longrightarrow P{-}CH_2CH{-}CH\overset{\cdot}{C}H_2$$
$$\qquad\qquad\qquad\quad |\quad\ |$$
$$\qquad\qquad\qquad\quad R\ \ R$$

(here P represents the rest of the polymer molecule).

Further gross structural irregularities can arise by radical abstraction of hydrogen atoms from polymer molecules, i.e. R'H above may be a polymer molecule; subsequent growth of polymer from the radical R'' leads then to a branched polymer molecule. A more subtle irregularity arises because the structural unit —CH$_2$—ĊHR— contains an asymmetric carbon atom (marked Ċ), and thus the groups attached to every alternate carbon atom along the polymer backbone chain may have either of two steric configurations. Steric regularity is of great importance in determining the conformation of a coiled or folded molecule and hence its ability to pack into an ordered (crystalline) state (see below).

Simple free radical initiation systems usually produce sterically irregular (random) structures, but in the 1950s new catalysts for ethenyl polymerisations were discovered by Natta and Ziegler which have the remarkable property of producing polymers with a more regular succession of steric configurations along the chain. An ethenyl polymer possessing steric regularity is known as a tactic polymer (from the Greek τακτικος, regular). If the substituents are all in the same steric order, the polymer is isotactic, and may be represented as

$$
\begin{array}{cccccc}
H & H & H & H & H & H \\
| & | & | & | & | & | \\
—C— & C— & C— & C— & C— & C— \\
| & | & | & | & | & | \\
H & R & H & R & H & R
\end{array}
$$

If the substituents are of regularly alternating steric order the polymer is syndiotactic, i.e.

$$
\begin{array}{cccccccc}
H & H & H & R & H & H & H & R \\
| & | & | & | & | & | & | & | \\
—C— & C— & C— & C— & C— & C— & C— & C— \\
| & | & | & | & | & | & | & | \\
H & R & H & H & H & R & H & H
\end{array}
$$

Ziegler–Natta catalysts function heterogeneously and commonly consist of the reaction product of an organometallic compound [e.g. triethylaluminium, $Al(C_2H_5)_3$] and a transition metal halide [e.g. titanium(IV) chloride, $TiCl_4$]. The mechanism of polymerisation on the surface of these catalysts is very complex and not understood in every detail, but is ionic in character and not free radical (see Chapter 9).

The polymerisation of ethenyl monomers, initiated and propagated by highly reactive free radicals or by ionic species, is classified as a chain-growth polymerisation. The polymerisation mechanism in which the monomer molecules can react with equal reactivity with either other monomer molecules or with polymer molecules, and which is exemplified by a polycondensation reaction, is classified as step-growth. Obviously, chance plays a great part in the growth of a polymer molecule. In the simplest terms, during the polymerisation of ethenyls (chain-growth) initiation by radicals is a random process and the polymer molecule grows in length by collision and reaction with monomer molecules, but if it meets the active end of another molecule growth may terminate for both. This is likely to be a rare event and most polymer molecules will be long. Collision with a transfer agent (the added species R′H) also causes termination of growth, and the higher the concentration of molecules which can act as transfer agents the shorter will be the average life time for growth of the polymer molecules.

In step-growth polymerisation monomer and polymer molecules of any size can react with each other at random and unlimited growth in size is prevented in practice by the eventual low concentration of active ends, and by the introduction of a monofunctional stabiliser, and in some cases by the reversibility of the reaction.

In either case all polymer molecules do not have the same size, but a statistical distribution of relative molecular masses is established and can be calculated if the mechanism is known, and measured in the laboratory. Therefore, when we speak of the relative molecular mass of a polymer sample we refer to some statistical average molecular mass. Usually this is the relative molecular mass averaged over the number of polymer molecules, i.e. the mass of the sample divided by the number of gramme molecules in it, and called the number average relative molecular mass $(\overline{M}_n)$. Some properties of polymers, especially the viscosities of melts and of concentrated solutions, are more sensitive to the larger molecules present in the distribution. That is, the viscosity is determined by both number and length of molecules. Thus, an average relative molecular mass which gives statistical weight to the length as well as the number of molecules is required to describe viscosities. The higher average commonly used is the weight average relative molecular mass $(\overline{M}_w)$ which can be measured experimentally. This is defined as:

$$\overline{M}_w = \sum_i M_i w_i / \sum_i w_i$$

where M_i is the relative molecular mass of polymer molecules of i monomer units long and w_i the weight fraction of this molecular species in the polymer.

The ratio of weight average to number average relative molecular mass gives some measure of the distribution of molecular sizes in the polymer. If all molecules are of equal size the ratio is unity. In a step-growth polymerisation when all functional groups have equal probability of reacting, $\overline{M}_w/\overline{M}_n = 2$. In some chain-growth polymerisations the distribution is very much wider and the ratio may be about 4. If branching reactions can occur the molecules have an additional means of growing in size and even broader distributions are produced.

POLYMERS AND FIBRES

In the following chapters of this book the relatively few major commercially established man-made fibres are briefly described. The

fibres are classified according to their basic chemical structures—
the sub-units or 'mers' which constitute the polymer molecules.
This is convenient because of the broad similarities underlying the
industrial production and the properties of the polymers within
each class. They are not listed by individual trade names because
fibres of the same chemical constitution are naturally given different
brand names according to their producer and, often, the country of
manufacture. This proliferation of names is confusing, and only a
limited number of relatively familiar fibres or historical 'firsts' will
be referred to by trade name.

A vast number of polymers have been made by chemists in research laboratories throughout the world, enormous numbers remain
to be made. Of this almost infinite variety of possible materials a
fairly large number could be processed into useful fibres. Why, then,
are there so few chemically distinct man-made fibres on the market?
Further, why is it that not all synthetic polymers can be fashioned
into practical textile fibres?

The answer to the first question lies in fairly obvious economic
and technical considerations. Before any fibre can find a market it
must be capable of production at a price acceptable to the buyer,
and sufficiently profitable to the seller. To attract the buyer the fibre
must offer some unique and advantageous combination of properties
which make it desirable in competition with established natural and
synthetic fibres.

The introduction of a fibre which is chemically new, as distinct
from an improved or modified existing one, is likely to involve very
heavy development costs and expenditure of capital on new plant
to produce chemical intermediates (monomer), plant to make polymer, and to spin and process the filaments. The new product must
be marketed and its uses developed in competition with cheap,
existing fibres. This is a venture not to be undertaken lightly, and at
the present time is possible only for large chemical and fibre-
producing companies, or for Government-backed organisations.
A very few new fibres are being introduced, and at the present state
of technical knowledge there are likely to be very few more. The
study and research does not stop, but few polymers can be made
from cheap and plentiful starting materials, carry the large capital
costs of production equipment and the costs of development, and
offer sufficiently novel properties as fibres to sell profitably to the
textile industry.

It is obviously more attractive commercially and technically to
modify and improve an existing fibre as much as possible and to
expand its sales by extension of its properties to adapt it to other
uses. Ideally, a fibre should satisfy a large number of different end-

uses so that eventually it may capture a large and broadly-based market. It should be versatile. Thus, several hundred variously modified forms of an established fibre may be offered for sale to the manufacturers of textile fabrics providing, for example, yarns of different numbers of filaments, different filament diameters and cross-sectional shapes, different opacities, textures, breaking strengths, extensibilities, and affinities for dyestuffs. Each of these yarns is designed to satisfy the requirements of processes of manufacture as well as the properties of specific ranges of many different textile products. It is important to realise that the processes of manufacturing finished goods may impose far more severe demands on the mechanical properties, quality, and chemical stability of a fibre (natural or synthetic) than does actual use or wear of the article. Yarns are expected to run virtually continuously at high speeds and under high and constant tensions over various rollers and guides: they have to withstand the sudden stresses of machine start-up, extremes of heat, and chemical treatments of dye and bleach. The introduction of an improved yarn may be dependent upon technical advances in the user industry, and the fibre producer has to be closely involved if he is to satisfy and keep pace with his customers' needs.

Nylon is a supreme example of a fibre whose properties have permitted exploitation in a vast range of products, and is adaptable to efficient, cheap processing on textile machinery.

Big, and growing markets mean larger volume production and cheaper fibre. Ultimately this situation rebounds on the happy fibre producer, and his profits decline, for the fibre becomes a 'commodity product', commonplace, over-produced and lacking in scarcity value, or novelty. His technical effort follows a cycle of evaluation and scale-up of manufacture of the new fibre; development of its uses and markets; improvements in quality; improvements in productivity; and, finally, as profitability is threatened, introduction of modified fibres with improved or quite novel properties. This driving need to maintain and improve profits to stay in business, spurred by competition, provides a guarantee of steadily developing quality and diversity of excellent products for the customer. But the necessary research and development is very expensive for the fibre producer.

FIBRE PROPERTIES

In order to understand why not all polymers make useful fibres, it is necessary to outline the general properties we require of a fibre, and to consider how these properties are related to the physics and chemistry of polymer molecules. If we consider a true textile fibre intended for broad application, not a speciality fibre with a specific limited use, then the fibre must be white or colourless, and able to be dyed to a wide range of shades with one or more established classes of dyestuff. The colours must be wash-fast and light-fast. The fibre must not be soluble in hot water nor in dry-cleaning solvents, and it must resist bleaches, alkalis, and hot, acidic dyebaths. The melting and softening temperatures will determine the heat limitations in processing and uses of the fibre; preferably the melting point should be above 200°C. The density should be low so that fine fabrics giving good cover at a low weight may be produced. The mechanical properties of the fibre are of major importance; it must have adequate tenacity (breaking load), extensibility, Young's modulus, elasticity, resilience, and abrasion resistance. Fabrics should exhibit good drape, crease resistance, be pleasing to handle and have a good appearance.

The numerical values for the various mechanical properties of fibres depend not only upon the chemical structure but also upon the physical structure, and, therefore, upon the details of the spinning process and the subsequent processing which the fibres have experienced. Thus, it is impossible to give a unique set of values of mechanical properties characteristic of each class of fibre, for each can be processed to give sets of values within a more or less limited range. This ability to produce a range of fibres and yarns from one polymer with pre-determined properties to suit different end-uses is, of course, one of the great advantages of man-made fibres.

Some idea of the typical ranges of the values of a selection of mechanical properties of several fibres is given in *Table 2.1*. The established older units of fibre science are the gramme and the denier: the denier, a term originated in the French silk industry, is the mass in grammes of 9000 m of yarn. The denier unit, therefore, gives a relative measure of the thickness of filaments of a given material, and is a useful technological unit, easy to measure accurately under factory conditions. A metric unit, the tex, is now replacing the denier system: one tex is the mass in grammes of 1000 m of yarn. Modulus and tenacity values are quoted in the literature in the older units of 'g per denier' or in new units of 'g per decitex' (1 dtex = mass of 10 000 m). Tenacity is thus not a true tensile strength or breaking load because the cross-sectional area of the fibre is not specified. The densities of organic fibres are similar (and close to unity) so meaningful working comparisons of tenacities (and of moduli) of different fibres can be made provided that the extensions to break are comparable. Extension modulus is usually expressed in terms of an extension of 100%: caution is needed in the use of

Table 2.1 SOME TYPICAL PROPERTY RANGES OF DRY MAN-MADE AND NATURAL FIBRES

Fibres	Specific gravity	Moisture regain (% at 65% RH and 20°C)	Tenacity (g dtex^{-1})	Extension at break (%)	Modulus of elasticity (g dtex^{-1})	Ratio wet tenacity to dry tenacity
Terylene						
Polyester, textile	1.38	0.4	3.6–4.5	30–15	90–104	1.0
Polyester, high tenacity	1.38	0.4	5·5–7.7	14–6	99–135	1.0
Nylon, textile	1.14	4.0	4.1–5.5	32–26	23–32	0.9
Nylon, high tenacity grades	1.14	4.0	6.3–8.1	22–14	36–45	0.9
Viscose, textile	1.52	c. 14	1.5–4.0	30–10	45–70	c. 0.6
Viscose, high tenacity grades	1.53	c. 12	4.0–10.0	10–5		c. 0.8
Acrylic	1.17	1.5	1.8–4.5	50–25	36–50	c. 0.8
Cotton	1.54	8	2.3–4.5	7–3	36–72	c. 1.2
Wool	1.32	c. 15	1.1–1.4	45–30	22–36	c. 0.8
Flax	1.54	12	4.5–6.3	2	c. 180	c. 1.2

this because it is determined from initial extension data, and stress–strain curves for textile fibres are decidedly non-linear. Non-textile fibres prepared from glass, metals, etc. (Chapter 11) usually have their mechanical properties expressed in true engineering or physical units. The SI units for modulus and tensile strength are N m^{-2}.

The effects on the physical properties of changes in temperature, and of water, are also of great importance. Textile processing and domestic laundering of fabrics demand ability to retain reasonable physical properties under hot, wet conditions: tyre cords function in a running tyre at temperatures well above room temperature.

The interaction between water and fibres is a very important aspect of the dimensional stability of fabrics, and of garment comfort in wear. A fibre which absorbs little or no water will show small or zero length changes when wetted, and will give fabrics of correspondingly good stability even under the extremes of humidity experienced in the British Isles. Drip-dry garments and fabrics can be made either from fibres with low water absorbancy or from fibres of higher water uptake (cotton, rayons) which have undergone hydrophobic surface treatments. Such treatments are rarely permanent and can adversely affect the handle of the fabric. However, it is uncomfortable to wear clothing which does not take up water from the skin on days of high humidity. The mass transport of water is mainly brought about by wicking along the yarns, so that the textile technologist can alleviate the problem to some extent by correct design of fabrics. The term 'moisture regain', or simply 'regain', is used to designate the mass percentage of water absorbed by a fibre in equilibrium with water vapour at a constant relative humidity.

FIBRE STRUCTURE

Although we are far from being able to make quantitative predictions of the actual mechanical properties of a polymer from the knowledge of its chemical structure alone, the research chemist making new polymers for fibres, or modifying existing ones, has a broad basis of theory and accumulated knowledge from which to work. We can go a good way towards guessing how a polymer will behave in general terms. Thus the synthesis of a polymer to achieve an approximation to desired fibre properties is a science rather than a hit-and-miss affair. But there are still plenty of surprises, for our knowledge of the detail of the organisation and interaction of molecules within any fibre is fairly limited. (And of course the obvious scientific way of meeting a commercial requirement is often too expensive.)

The properties of a given fibre can be described in terms of a simplified molecular model; as more facts become known the model is either refined or rejected. Each fibre requires its own model, and we cannot know how well or badly the model represents reality; we only know how well it fits our interpretation of experimental observations. In considering the basic properties of fibres we have to examine four levels of organisation of structure which contribute. First there is the chemical composition and structure of the repeat units in the polymer molecule, and the properties of the chemical bonds which constitute the 'backbone links' of the molecule. Secondly, the sizes and shapes of the polymer molecules as a whole, and how they interact with each other. must be considered. At a higher level comes the three-dimensional arrangements of the molecules into some kinds of aggregates, and the interactions and spatial array of these aggregates to form the bulk fibre. Finally, the surface of the fibre has to be considered, for frictional and tactile properties, soiling, static electrical charging, wetting, and laundering are all important aspects of the surface of textile fibres. (One natural fibre showing gross physical differences between surface and core is wool, which has a surface structure made up of flat, overlapping cells so that is has a scaly appearance and character—see *Figure 2.1*.)

Models for the structure of a fibre must include dynamic mechanisms for response to loading, deformation, creep, fracture, etc. A

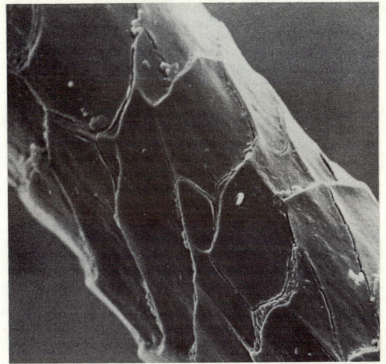

Figure 2.1   Stereoscan photograph of a crossbred wool fibre, magnification × 3760

detailed discussion of all of these features of fibre structure and properties, and of the theories and models proposed to explain them, is beyond the scope of this introductory book. A greatly simplified picture is presented here, with due caution of its many limitations.

The forces between polymer molecules are determined by the chemical structure of the repeating unit. These forces, which are of the same nature as those between small molecules, are potentially strongest in polymers which contain dipolar, ionic, or hydrogen-bonding groups. The full strength of these directional forces can be realised if the regularity of structure and flexibility of the polymer molecules permit the correct correspondence and alignment of the interacting groups with each other, and if bulky side-chains do not prevent sufficiently close approach. Whilst the forces of attraction between such interacting groups individually may be small, the cumulative effect of a large number of groups, joined together by covalent bonds to form the polymer molecule, is large. Thus, the useful physical properties of polymers are dependent upon the degree of polymerisation and are not fully developed until fairly high relative molecular masses are attained.

The development of an ordered, close-packed structure within a polymer sample constitutes crystallisation. No regular crystal faces, characteristic of simple crystalline solids can be discerned, but relatively sharp X-ray diffraction patterns can be obtained from crystalline polymers, similar to those from powdered crystalline solids. The size of these crystalline regions within the polymer may be only about 10 nm. Polymer molecules (100 nm–10 μm long) may pass from a crystalline to a non-crystalline (amorphous) region, and may also pass through into a second (and more) crystalline region. Thus, we develop a picture of small, ordered, crystalline zones which are separated by less well ordered, amorphous regions, but which are all connected together by a random network of polymer molecules.

Perfect crystalline order within a polymer sample is prevented by the restricted mobility and entanglements of the long molecules, and by chemical structural imperfections. Under special conditions of crystallisation, e.g. from dilute solution, or by slow cooling of a melt, less imperfect crystalline structures can be grown: indeed, polymer single crystals can be grown from a very dilute solution, and the structure of these is referred to later.

The structural requirements for crystallisation are especially relevant for hydrocarbon polymers where only the weaker shorter range van der Waals forces are operative: linear polyethene is crystalline, the molecules can pack closely together. If bulky hydrocarbon side-groups are introduced, less crystalline, or wholly amorphous, polymers are obtained. However, highly crystalline polymers may be prepared by the use of the stereo-specific catalysts discussed above. The feature of these polymers is their high degree of structural regularity: the steric hindrance of the adjacent bulky side-groups of isotactic polymers is relieved by the assumption of a regular helical conformation by the polymer molecules.

By no means all polymers have a sufficiently regular or flexible structure to permit the development of crystalline regions. Polymers which are devoid of crystalline structure (amorphous polymers) in the sense described above can be used to make fibres (see 'acrylic' and 'vinyl' fibres, Chapter 8). Less is known about molecular arrangements in amorphous polymers, and in the amorphous regions of crystalline polymers, because of the limited experimental techniques available to study them.

Polymer single crystals grown slowly from very dilute solutions have the appearance of very thin (c. 10 nm), flat platelets (lamellae). Study of these crystals gave the surprising observation that the long axes of the polymer molecules are aligned perpendicular to the plane of the lamella. Because the single crystal is so thin, this means that the long polymer molecules must fold back on themselves and pack into a structure something like a jumping cracker (*Figure 2.2*). The discovery and detailed study of these intriguing folded structures dates only from the late 1950s. It is not yet clear to what extent chain folding occurs during polymer crystallisation from concentrated solutions and melts, but there is good evidence for the presence of folded chain crystallites in some melt crystallised polymers, and it is likely to be a general feature. In the extreme view, the crystalline polymer structure can be regarded as comprising perhaps imperfect

Crystal face

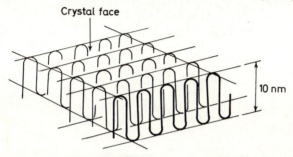

10 nm

Figure 2.2   Diagram illustrating idealised chain-folding and molecular packing in polymer single crystals (not to scale, but the thickness would be about 10 nm, as indicated)

folded-chain crystalline regions (built up from lamellae stacked face-to-face) tied together by a network of disordered polymer molecules. The imperfections in the crystal lattice, ends of molecules, boundaries between the stacks of lamellae, and the intermeshing connecting molecules (tie molecules) may be taken together as 'amorphous regions'. *Figure 2.3* attempts to give a diagrammatic picture of a possible structure.

The ability of a molecule to fold is clearly dependent upon its flexibility and hence chemical structure. Some very stiff molecules can be conceived which would not fold, or fold only with difficulty, and these may be expected to crystallise (if at all) in an extended molecular conformation like rods. Such polymers offer the prospect of quite different mechanical properties and are the object of some interest at present (Chapter 6, page 80, for example): but stiff molecules introduce fabrication problems because the property contributes to difficult solubility, very high melting points or infusibility, and high melt viscosities. Under conditions of extreme pressure and shear it is possible to force some conventional polymers to crystallise in an extended-chain form, rather than as folded-chains. However, reversion to the folded structure is observed on heating, and so far

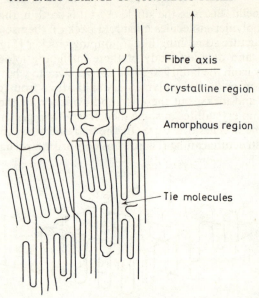

Fibre axis

Crystalline region

Amorphous region

Tie molecules

Figure 2.3   Diagrammatic illustration (two dimensional) of possible molecular organisation of a drawn crystalline fibre such as nylon

there seem no realistic commercial possibilities for making these materials.

Examination of films of some crystalline polymers under the optical microscope reveals the presence of much larger features than the crystalline structures discussed above. The commonly observed entities are spherulites, *Figure 2.4* (often seen in melt crystallisations of simple organic compounds), which are birefringent, spherically symmetrical structures formed of aggregates of the small crystallites. They are not single crystals. Spherulites are nucleated structures, that is, they grow upon a nucleus which may be a residual unmelted fragment of a crystallite or a particle of some suitable impurity, so that their sizes and numbers are determined by the previous treatment, purity, and crystallisation conditions of the polymer. Control of the growth of spherulites is vitally important in melt-spun synthetic fibres as the presence of large entities can make fibres brittle, and affect their reflectance of light and hence appearance. This control is normally achieved by adjustment of the spinning conditions, but nucleating agents are known and described in the patent and scientific literature.

An appreciation of the conditions under which a polymer will crystallise and the rate at which it will crystallise is of major impor-

Figure 2.4   A spherulite grown in a thin film of 6.6 nylon at 260–265°C and viewed between crossed polars on a microscope: the diameter is about 150 µm

tance in fabrication of articles and fibres from a polymer melt. If a molten polymer is rapidly cooled it can be taken to a temperature which is too low for crystallisation to occur at any detectable rate; the solid is amorphous. Crystallisation is prevented by the restrictions on mobility of the long molecules in the solid state. Rapid cooling of the filaments is a feature of melt spinning processes (Chapter 3). As the temperature is raised, and thermal energy permits greater molecular mobility, then crystallisation can eventually begin. The room temperature crystallisation of polyamides especially is promoted by water, which probably reduces molecular interaction and hence increases mobility. The dimensional changes brought about by uptake of water, and consequent swelling or crystallisation effects, can play havoc with the stability of the many miles of freshly spun yarn wound up on a cylinder. The spinning process has to allow for the control of these phenomena.

The concept of different degrees of molecular mobility within a solid polymer is fundamental to our interpretation of the mechanical properties. The accepted 'melting point' of a crystalline polymer can be defined and observed experimentally as the temperature at which the crystalline features melt. It is not sharp, because of the imperfections of the crystallinity, but normally coincides well enough with the range of temperatures within which bulk viscous flow of the polymer is observed. An amorphous polymer has no such defined melting point, but will show a transition from a brittle glass-like solid via a leathery or rubbery state to a viscous liquid over a much wider range of temperature. The 'crystalline' polymer has both amorphous and crystalline character: the crystallites are regions of high molecular order, in the amorphous regions the molecules or parts of molecules are much less constrained, hence less thermal energy is required to permit the onset of molecular chain mobility within the amorphous regions. Only at higher temperatures will there be sufficient energy to disrupt the crystallites.

Thus, there is a second temperature range characteristic of the non-crystalline regions within a crystalline polymer, and characteristic of the onset of longish range molecular chain mobility within a wholly amorphous polymer. This temperature is a transition temperature; distinct from a melting point, it does not measure a discontinuous change in structure, and is referred to as the 'glass transition temperature' (T_g) marking the transition from glassy to rubbery state in an amorphous polymer. The glass transition temperature is not well defined, and is not uniquely determined experimentally because the detection of mobility implies a response to the time scale of the detecting system.

Detailed study of polymers reveals the presence of other minor 'transition temperatures', each probably representing some degree of freedom of molecular movement; in a useful fibre-forming polymer the whole transition region may extend over two hundred degrees of the temperature scale. Study and interpretation of the transitions and relaxation processes in molecular terms is an active field of research.

One additional, imposed, molecular ordering process is vital to the production of fibres. The irreversible stretching and drawing stages which are associated with the various spinning technologies (described in Chapter 3) produce a preferential orientation of the long axes of the polymer molecules along the direction of the fibre.

Orientation produces an anisotropic fibre with greatly enhanced mechanical properties in the direction of its length. Molecular orientation occurs whether the fibre-forming polymer is crystalline to some degree, or not at all. Indeed, unless a polymer can be oriented in this way it is unlikely to display good fibre properties.

Orientation by stretching introduces some degree of molecular order to an amorphous polymer, and develops fibre properties, but it does not necessarily bring about crystallisation.

The structure of a crystalline polymer is changed by the orientation process; of necessity molecules and blocks of molecules must move as the filament is stretched. The length of the fibre may be increased by several hundred per cent, so fairly dramatic moves are involved. Drawn synthetic and stretched regenerated fibres generally show a parallel, fibrillar fine structure when examined by electron microscopy. Possibly slip occurs at crystal planes, and some folded chains are extended, but no clear picture of these important processes has yet emerged.

Before synthetic fibres can be drawn at industrial processing speeds it is essential to heat them to, or above, the glass transition temperature, to obtain a sufficient degree of molecular flexibility to permit orientation. At the high speeds of drawing, undrawn fibres below T_g tend to be brittle. Because of the work done on the fibre at drawing, heat is generated.

Natural vegetable and animal fibres, as may be expected, have highly organised internal structures, and are cellular materials. Plant fibre cells are highly elongated and fibrous in form; the cotton hair is a single collapsed cell, the walls of which are built up from bundles of parallel microfibrils of cellulose which spiral around the cell. Wool and hair have a most complex, multi-cell structure, and the differentiation of core and surface structure mentioned above; silk, as an extruded fibre, has no cellular structure. The chemical structures of the proteins of silk and wool are complex and fascinating, they are considered further in Chapter 4.

FIBRE STRUCTURE AND PROPERTIES

In very crude terms the simple mechanical properties of fibres and plastics can be explained by the flexibility and movement of the molecular chains in the amorphous regions. Hence the basic importance of the transition temperature range and the glass transition temperature.

Within and above the transition temperature range the flexible amorphous regions transmit imposed stresses throughout the material, permit the relief of strains, and the absorption and dissipation of mechanical energy. Thus, they provide toughness and flexibility. Extensibility (reversible) is provided by the uncoiling of the molecules in the amorphous regions: recovery forces (largely entropic) are provided by the thermal energy of the molecules. The modulus of extensibility is thus determined by molecular flexibility within the amorphous regions, and is high at low temperatures falling rapidly as the glass temperature is passed through and remaining low at higher temperatures. If a stress is applied very rapidly to a fibre the molecular mobility may not be sufficiently great to permit fast enough response and it will undergo glassy fracture—in effect T_g is raised to a very high value because of the very short time scale. As the other extreme, the sustained application of a load for a prolonged time will result in molecular flow and a permanent deformation. The position of the glass temperature on the temperature scale relative to room temperature has a profound effect on the observed everyday properties of a fibre.

Water can profoundly reduce the glass transition temperature, by solvating the molecules and decreasing forces of interaction, penetrating and swelling a hydrophilic polymer. The increased molecular mobility may permit the release of strains previously imposed on a fibre at spinning and processing with a resulting shrinkage in length.

Crystallisation provides a device for locking molecules together, effective to a higher temperature. The crystalline regions act as reinforcing zones to increase the stiffness of the material, and prevent molecular flow above the glass temperature. In a rather poor analogy, they provide the bones and anchor points on which the amorphous muscles can act. Crystallisation also decreases the penetration and swelling of the fibre by water (and other liquids), improving the wet physical properties.

Orientation by drawing or stretching increases the modulus and the strength of a fibre, but reduces the extensibility. Probably the molecules in the amorphous regions are pulled into new, more strained conformations with less potential reversible extensibility and less mobility. In a crystalline polymer orientation is stabilised, and 'set in' by crystallisation. Fibres of different tenacities and extensibilities can be obtained by control of the degree of orientation.

It is important to realise that synthetic fibres are in metastable states because of strains imposed at spinning, drawing, etc., and will relax to states of less strain if permitted, with shrinkage. The effect of water has been mentioned, obviously relaxation can also be effected by heat treatments. Heat treatments can also be used to deform a fibre or fabric and set it in the new deformed state. The mechanism of heat setting will be dependent upon the T_g of the fibre and its crystallisability. A high T_g fibre can be deformed and 'locked' in position by heating above T_g, deforming, and cooling in the new position. Adjustment of crystalline structure or further crystallisation will help stabilise the new configuration. Pleats can be set in fabrics made from such fibres, and will remain for a long time unless the fabric is reheated to the pleating temperature, or unless the T_g is sensitive to water. These fabrics are of course susceptible to accidental creasing and wrinkling under hot and wet conditions—at the laundry—but are crease-resistant under conditions of wear.

If the T_g is below room temperature there is no mechanism for retention of deformation other than by adjustment of crystallisation, and fabrics will not retain pleats well.

'Terylene' polyester fibre is an excellent example of a high T_g material (T_g about 90°C), little affected by water, and used in suitings to exploit the pleating, crease-resistant property (the crease stays in the trousers—even in the wet!). Cellulosic fibres have much higher glass temperatures but are very sensitive to water unless highly crystalline. Nylon has a lower T_g (about 50°C when dry) which is also sensitive to water, falling to about room temperature, but it can be heat set to give dimensionally stable fabrics, probably held stable by crystallisation and interaction of the highly polar amide groups. Such diverse behaviour illustrates the complexity of the subject and the danger of generalisation. But perhaps the mysteries of steam ironing and some advertising are more clear!

Searching for improved fibres then, the polymer chemist looks for structures which have high symmetry and regularity, which have polar groups, and which have stiffness (and are cheap).

Molecular flexibility arises by rotations about the covalent bonds in the polymer backbone, therefore to obtain stiff molecules there must be steric or other barriers to rotation, or the number of rotations should be reduced—this can be achieved by replacing chains of carbon atoms by rings of carbon atoms. The effect of a ring of carbon atoms is well illustrated by the comparative melting points and glass temperatures of the two polyesters:

$$\sim\!\!\sim\!OCH_2CH_2O.CO\,(CH_2)_4\,CO\!\sim\!\!\sim \quad mp \quad 50\ ^\circ C \ T_g \sim\!-50\ ^\circ C$$

$$\sim\!\!\sim\!OCH_2CH_2O.CO\!\!\left\langle\bigcirc\right\rangle\!\!CO\!\sim\!\!\sim \quad mp \quad 262\ ^\circ C \ T_g \sim\!90\ ^\circ C$$

The effect of symmetry can also be seen by comparing the melting point and T_g of the following polyester with the above:

$$\sim\sim OCH_2 CH_2 O.CO \text{—(benzene ring, meta)—} CO \sim\sim$$

mp 240 °C
$T_g \sim 50$ °C

The effect of directional forces can be shown by comparing polyesters and polyamides:

$$\sim\sim O(CH_2)_6 O.CO(CH_2)_4 CO \sim\sim \qquad \text{mp } 56°C$$

$$\sim\sim NH(CH_2)_6 NH.CO(CH_2)_4 CO\sim \qquad \text{mp } 265°C$$

strong dipolar interaction or hydrogen bonding between the carbonyl $C{=}O$ and the N—H groups gives rise to the increased rigidity of the polyamide molecules in comparison with the polyester.

Rings have similar effects in polyamides:

$$\sim\sim NH(CH_2)_6 NH.CO \text{—(ring)—} CO\sim\sim \qquad \text{mp}\sim350°C$$

and symmetry:

$$\sim\sim NH(CH_2)_6 NH.CO \text{—(benzene ring, meta)—} CO\sim \qquad \text{mp } c.200°C$$

By destroying the dipolar interaction it is possible to produce low melting polyamides which are rubbery and non-crystalline. This involves the substitution of a bulky group for the hydrogen of the amide group:

$$\sim\sim \underset{\underset{CH_3}{|}}{N} (CH_2)_{10} CO \sim\sim \qquad \text{mp}\sim60°C$$

Many other examples and more detailed examination of these structural effects are to be found in polymer chemistry textbooks.

By incorporating such structural features the chemist can obtain high melting, high T_g, crystalline, high modulus materials. Alternatively, he can reduce stiffness by building-in flexible units, and in the ultimate produce rubbery materials (Chapter 10).

Having selected his main structural chemistry, the organic chemist may wish to bring about minor modifications, usually by the copolymerisation of small amounts of an additional monomer, to facilitate ease of processing, to increase the ease of dyeing, and to extend the range of dyes which can be applied to the fibre. Working closely together the organic chemist, physical chemist, physicist, engineer, and textile technologists can study and modify crystallisa-tion, orientation, properties, and processes to obtain the full potential of a fibre for a variety of end-uses.

Before taking up in more detail the chemistry of commercial fibres, the basic spinning methods will be described.

3

Spinning Man-Made Fibres

The general methods for the conversion of organic polymers to fibres—the spinning technologies—are considered in this chapter, before describing the production of the individual polymers. Whilst this is not the chronological sequence of events, an early account of spinning methods will avoid undue repetition and forward reference.

The essential step of any spinning technique is the transformation of a solid polymer into a liquid state so that it may be extruded through fine holes, and the emergent jets then returned without loss of form to a coherent solid state. A practical spinning technology and machine must be sufficiently flexible in operation to allow the efficient and economic production of a wide range of yarns of high quality and uniformity. By a wide range of yarns is meant a variety of filament numbers, diameters, and physical properties from one given polymer type. A machine is not necessarily required to be able to spin two or more chemically different polymers.

MELT SPINNING

Conceptually, melt spinning is the simplest technique: a thermoplastic polymer is melted, the molten, viscous fluid is extruded, and emerging filaments are solidified by cooling. (The term 'thermoplastic' applied to a material means that it is softened by heat so that it can be shaped, and hardens again on cooling, the change being a physical change, so that it may be repeated indefinitely.)

It is easy to demonstrate in a school laboratory the formation of fibres from a melt. Small pieces of nylon polymer can be melted in a hard glass test-tube over a small gas flame. Oxygen will badly degrade the polymer, but can easily be excluded by first adding sufficient water to cover the pieces and gently boiling this away to displace the air with steam before heating more strongly to melt the polymer. If the melt is touched with a cold glass rod, and then the rod withdrawn, a filament can be pulled out. Fibres can be obtained in a similar way from pieces of polyethene. Scrap plastics (correctly identified) is a possible source of polymer, but it is better to buy granules from a supplier of laboratory chemicals.

In principle, melt spinning can be used for any polymer which has its melting point below the temperatures at which the rates of thermal decomposition reactions become appreciable. This temperature gap needs to be not less than about 30°C to provide a good working range of melt viscosities, and so that molten polymer may be kept in the spinning machine for some reasonable time without harmful degradation. Polymer melts are very viscous liquids, the viscosity decreases greatly as the temperature is raised. The time for which a polymer is molten in the machine depends upon the spinning process, i.e. the way in which the machine has to be run to make a desired yarn. But, although the design will be such that almost all molten polymer will pass through with a residence time of perhaps no more than ten minutes, very small quantities of polymer will be held up for much longer times in slowly moving regions, and in static pockets, or will be recirculated by pumps. Detrimental chemical changes may occur over these longer times at high temperature. These chemical changes may show themselves in many subtle ways in poor and variable yarn physical properties, and in changes in the dyeing behaviour. Good engineering design of the melting device is very important to minimise static zones in the flow path of the melt.

Two extremes of thermal degradation behaviour are illustrated by nylon (the type known as 6.6 nylon, see Chapter 6) which is slowly converted to an infusible, insoluble material, and polyester (see Chapter 7) which breaks down to a lower relative molecular mass and lower melt viscosity polymer.

The simplest technique of melt spinning is outlined in *Figure 3.1*. The solid polymer granules are fed from a hopper to the melter, and molten polymer collects in a melt pool from which it is pumped at a constant rate through a filter pack to the spinneret. The extruding filaments are cooled by an air blast, and solidify, and are then brought into contact with guides and wound up. We shall consider each of these steps in more detail.

The solid polymer is supplied to the hopper in a form known as 'chip'—pieces about the size of rice grains which will flow through the hopper and melt easily. Large pieces are slow to melt because heat transfer through viscous molten polymer is poor. If melting is incomplete, undesirable large crystallites, which are nucleated by crystalline residues from the initial solid, may be formed in the spun yarns. The chip either flows by gravity, or more usually is fed by a screw, from the hopper to the melting zone.

Two basic types of melter are in use, the grid melter and the screw

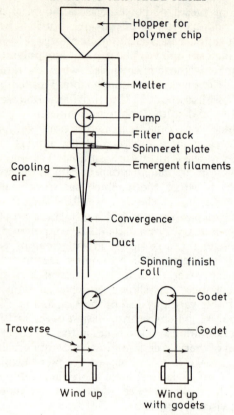

Figure 3.1   Diagrammatic representation of melt spinning

extruder. A simplified diagram of a grid melter is given in *Figure 3.2*,
and a photograph of the grid (seen from above) is reproduced in
Figure 3.3. The chip is fed down on to a grid of hollow metal fins
which are heated internally by the vapour of an organic liquid
(usually an azeotropic mixture of diphenyl and phenoxybenzene).
Electrically heated grids are used, but vapour heating has the
advantage of avoiding localised overheating. Temperature can be
controlled by adjusting the pressure above the boiling liquid, and
vapour is piped to melters from a central boiler.

The polymer melts at the surface of the grid and flows through
into the cavity beneath—the melt pool. An inert atmosphere has to
be provided to prevent oxidation of the polymer; this is usually
nitrogen, but may be steam for nylon (Chapter 6). If the level of

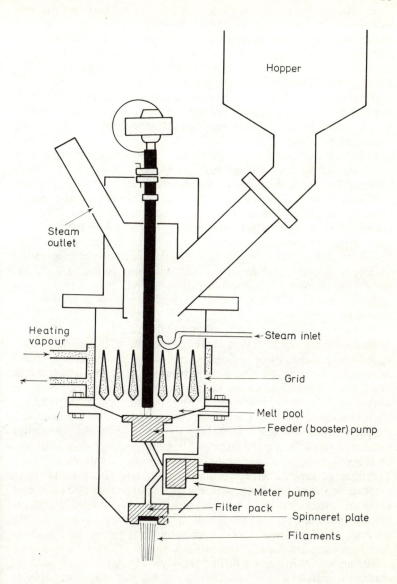

Figure 3.2 Grid melter as used for 6.6 nylon (gravity feed)

molten polymer rises above the grid it prevents further melting, and so the grid melter is self-regulating to some extent, but there is a danger that an excess of molten polymer above the grid may cool and solidify with the whole mass of chip, and nowadays level control devices are usually fitted to regulate the delivery of chip. These control devices serve the purpose of keeping the time in the molten state to a necessary minimum, and more constant. Entrapped gas bubbles can escape from the melt in the pool, and it provides a holding-time to enable complete melting of crystal nuclei and bits of unmelted solid.

The molten polymer is pumped from the pool to metering pumps which deliver polymer at a carefully controlled rate to the spinnerets. The melt is overfed to the meter pumps so that they are never starved, and the excess circulates back to the pool. Gear pumps are used to forward and meter the melt. A precise volume of liquid is carried between the carefully machined teeth of a gear wheel rotating in a close fitting chamber. A second gear wheel intermeshes with the first. Liquid flows into the gap as the teeth separate, is carried round for one revolution, and displaced (perpendicular to the plane of the wheels) as the teeth intermesh again. Gear pumps are robust and accurate, working at 300°C and higher. A simple gear pump is shown in *Figure 3.4*.

Between meter pump and spinneret plate is a filter pack. The filter is usually made up of layers of graded alumina, or purified sand, supported between discs of fine metal gauze. Sintered porous metal plates may be used also. The filter removes particles which might block the spinneret holes, or pass through to give weak points or lumps in the filaments and bring about faults and breaks during later processing or use. These particles are likely to be bits of un-melted polymer, and degraded solid polymer arising within the melter itself or from the polymerisation process, and aggregates of additives to the polymer [most commonly titanium(IV) oxide delustrant]. The filter becomes obstructed after some days, with a build up in pressure at the meter pump, and, therefore, filter pack and spinneret plate are assembled in one block which can be taken out and re-placed without serious interruption of spinning. Good quality polymer ensures fewer interruptions of this nature and smooth running of the machine.

The spinneret holes are drilled through a metal plate, usually of corrosion resistant stainless steel or nickel alloy, and capable of withstanding high pressures. The holes are arranged in precise patterns, often along the circumferences of concentric circles, to allow uniform cooling of the emergent filaments (illustrated by Figure 3.5). The capillary, the actual spinneret hole, may be a few

Figure 3.3   View from above of a simple nylon melter, showing the grid, steam inlet, and connections to the vapour heating system

Figure 3.4   A simple gear pump, split to show the working parts

millimetres long and 0.1–0.8 mm in diameter. The thickness of the plate above each hole is drilled out and channelled to provide uniform access of molten polymer to every capillary. The holes must be precisely identical. Spinnerets are expensive!

Non-circular holes are also used to make filaments of various cross-sectional shapes. The filament does not retain the precise shape of the non-circular hole from which it emerges, for surface tension and the tension drawing it away from the spinneret tend to restore it to a circular section. Shape retention is better if the polymer melt is more viscous and is cooled more rapidly as it emerges. As examples of the large number of differently shaped holes which have been patented, a Y-shaped hole will yield a three-lobed filament

Figure 3.5   Nylon filaments extruding from a spinneret

cross-section, which may be almost triangular, and a C-shaped hole can produce a hollow filament with an O-section. Non-circular fibres have special applications, they are not simply curiosities, and we shall meet up with them later.

Grid melters were introduced by Du Pont for the melt spinning of nylon, nowadays screw extruders are also used as melters for fibre spinning, and indeed are essential to handle the high viscosity melts of polyalkenes. Gravity flow of high viscosity melts is too slow to permit use of a grid melter. Polymer chip is fed from a hopper into the flights of the rotating screw which is contained in a barrel and which may be mounted either horizontally or vertically. The chip is carried forwards (or downwards if the screw is mounted vertically) by the screw into a melting zone. The barrel is heated by electrical heaters. By changing the profile of the screw, the melting solid is compressed, and the melt carried on and fed into a melt pool at the end of the screw. The rest of the process is then essentially as before. The screw extruder combines the steps of conveying solid chip, melting, and forwarding the melt, into one device. A suitably designed extruder can have a very high melting capacity and a very high throughput of polymer—perhaps 50 or 100 kg/h^{-1} which makes a staggering number of kilometres of fine filament!

Both grid melters and screw extruders supply molten polymer to more than one meter pump/filter pack/spinneret assembly. Molten polymer may also be provided directly from a continuous polymerisation reactor and fed to many spinning positions. In all cases the rate of provision of polymer melt has to be matched to the rate of conversion to yarn.

The diameter (the tex as defined in Chapter 2) of a spun filament is not determined by the size of hole from which it emerges, but by the rate at which polymer is being supplied to the hole and by the linear velocity at which the filament is being wound up. The wind up speed is very high, thus, whilst in the molten state the filaments are being accelerated away from the spinnerets and reduced in diameter. An air blast below the spinneret plate cools the filaments, and at some point they solidify and then proceed at almost constant diameter. The control of filament diameter is vital, and hence the importance of maintaining constant and uniform rate of feed to each spinneret hole, equal hole dimensions, and the same cooling conditions for all filaments from the spinneret. Wind up speeds of up to two thousand metres per minute are used, and solidification takes place in a few tenths of a second and within a metre of the spinneret plate.

The solidified filaments can be brought together without fusion and into contact with ceramic guides. The bundle of filaments may be split up on these guides into smaller groups of fewer filaments, each to be wound up separately and processed into yarns. The whole bundle is passed through a vertical duct down to the wind up mechanism. Passage through this duct maintains uniform cooling conditions and provides protection. Further physical changes may occur, depending upon the polymer being spun. The filaments are super-cooled and solidified so quickly that little crystallisation has time to take place, and the polymer has largely the amorphous structure of the melt itself. Some orientation of the molecules is brought about by shearing forces as the filament is accelerating and drawing down in diameter. 6.6 nylon filaments are exposed to steam within the duct, which takes the form of a tube and is called a 'conditioner'. Less than half a per cent of water may be taken up by the dry nylon filaments, but this is sufficient to initiate some further rapid crystallisation. This crystallisation and water absorption stabilises the filaments when they are wound up. Absorption of water can cause an increase in length of the filaments, because they swell, or it may relieve stresses in a filament wound up at too high a speed causing it to shrink. The additional crystallisation which takes place leads also to an increase in length. The molecules crystallise most easily upon the existing small crystalline regions which act as nuclei. These are oriented to some degree, and further growth upon them induces orientation in the previously random structure and causes an increase in length.

If these changes in length took place on the package after the filaments had been wound up, the whole would be loose and liable to slough off. The fractional change in length may be very small, but there are many miles on the package!

The changes in crystallinity may be from about 20% crystalline in the freshly spun yarn to 35% crystalline in the 'conditioned' state. Subsequent drawing of the nylon to give high orientation (see below) might increase the crystallinity to about 40%.

Before winding up the filaments, spinning finish is applied by contacting them with the surface of a rotating wet roller, or by a capillary feed device. The quantity of liquid put on to the filaments is carefully controlled. Spinning finish is a complex preparation of oil lubricants emulsified in water; wetting agents, antistatic agents, and adhesives may be added also. Its purpose is to provide yarn lubrication for subsequent processing over guides and rollers, and to disperse static electrical charges during processing which would otherwise cause mutual repulsion and separation of filaments and consequent snagging. The formulation of finishes depends upon the yarns being spun and the subsequent processes to which they will be exposed. Ultimately it is washed from the fabric.

The wind up mechanism must run at a strictly constant linear yarn speed, and provide a stable, low tension package from which the yarn can be unwound evenly. The filament bundles may be stabilised by passing round two rotating rollers known as godets (*Figure 3.1*, and *Figure 3.6*), and are then wound on a cylinder which is perhaps

Figure 3.6   Nylon filaments being wound up on a simple experimental spinning machine

about 15 cm in diameter. The cylinder is surface driven to maintain constant linear wind up speed as the diameter increases. To obtain a cylindrical, stable build of filaments, the threadline is traversed to and fro across the cylinder by a guide. The traverse is programmed by a cam, and a great deal of skill and thought goes into its design, otherwise a hopelessly collapsed, entangled mess can result! The photo-graph shows a single wind up position and single bundle of filaments. Commercial machines have many such positions side by side, probably with multiple threadlines being wound up at each—it depends upon the yarns being produced. In addition, the machine will be double-sided. The spinnerets are on the floor above, at about head height for easy working and to allow access to the filament cooling system below them. The next floor up houses the melters and feed hoppers. A plant producing fibres has to be scrupulously clean—customers do not want grubby yarns!

When spinning fibre to be turned into staple, the filament bundles from several positions may be brought together for collection as a thick bundle known as 'tow'. Fine quality control of individual filaments is less important than for continuous filament yarns, and the aim is for high productivity. Subsequent processing to staple fibre is described in Chapter 12.

Melt spun filaments have very high extensibility (*Figure 3.7*). They are almost isotropic in physical properties, with little orientation of

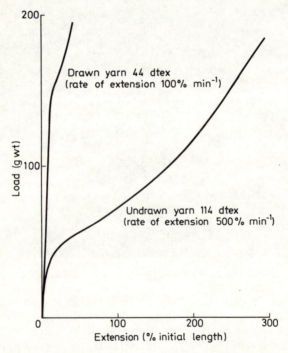

Figure 3.7   Load-extension plots for corresponding undrawn and drawn 6.6 nylon yarns

the molecules along the fibre direction. The high extensibility of the spun filaments is not reversible, and does not occur by a uniform reduction in diameter over the whole length of the elongating fibre. As the filament is extended a yield point is reached (at about 10% extension) and from then on the reduction in diameter becomes localised at a point, known as a 'neck', and is irreversible. This process is known as 'drawing'. It is easily observed, and best appreciated, by slowly hand stretching narrow strips of polyethene sheet —cut from a 'polythene' bag—or a crude fibre pulled from a melt as described above. (The improvement in physical properties after drawing will be obvious.)

Drawing produces molecular orientation along the fibre axis, as discussed in Chapter 2. Draw ratio is defined as the ratio of the drawn length to the undrawn length of the yarn. The higher the draw ratio, the greater is the degree of orientation, and the higher is the tenacity and modulus (and the lower is the extensibility) of the yarn. Thus, high tenacity yarns for tyre cords are drawn to a high draw ratio (about 5 for nylon), whilst textile yarns are drawn to a lower draw ratio (about 3), but the degree of orientation in the drawn yarn depends upon the orientation put in at spinning (i.e. spinning speed) as well as upon the imposed draw ratio. To obtain the molecular mobility necessary to allow high speed drawing, the polymer must be above its glass transition temperature (Chapter 2). It is advantageous to heat all fibres when drawing to maximum draw ratios.

Yarns are drawn continuously by running between two rollers, the second of which (the draw roll) is rotating at a faster surface speed than the first (the feed roll). A few wraps round each roll provides grip. The relative speeds of the rolls are controlled and can be varied to produce different draw ratios. The neck is localised, and draw controlled, by wrapping the yarn round a metal or ceramic pin inserted between the rolls (*Figure 3.8*). The yarn is heated either by a heated feed roll, or by running over the surface of a hot plate between the rolls.

Melt spinning and drawing have been described as two separate stages with an intermediate wind up on a package and transfer to a drawing machine which may be in a different part of the factory, and indeed this is the common practice. However, the two can be combined into one continuous operation ('spin-draw'), but the problem of matching the linear wind up speed after drawing to the already high speed of melt spinning imposes some difficulties and limitations.

The drawn filaments are wound up on a cylinder, perhaps 5 cm in diameter, and may now be described as continuous filament yarn, for in some cases this is the end product. The filaments are twisted together at wind up by passing through a loop of wire which travels

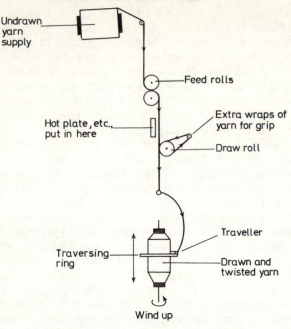

Figure 3.8   Drawing yarn

freely in a circular track around the rotating cylinder. The loop is dragged round by the filaments, and twists the free length of yarn which precedes it (*Figure 3.8*). Alternatively, the yarn may run over the edge of a cap above the wind up cylinder.

DRY SPINNING

Many polymers cannot be melt spun because they are decomposed by heat before, or close to, the melting temperature. These polymers may be spun from solution. Two methods are in use, and are known as 'dry' and 'wet' spinning.

In the dry spinning process a concentrated polymer solution is extruded through spinneret holes into a current of hot gas which evaporates the solvent (*Figure 3.9*).

The solution should be highly concentrated to minimise the quantity of solvent which has to be vaporised and recovered, but there are practical limits to the concentrations of solutions which

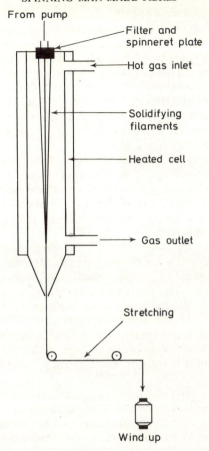

From pump

Filter and spinneret plate

Hot gas inlet

Solidifying filaments

Heated cell

Gas outlet

Stretching

Wind up

Figure 3.9   Diagrammatic representation of dry spinning

can be prepared and spun. A solution which is too concentrated may have highly visco-elastic, jelly-like properties and be difficult to pump. The extruding filaments may be unstable and break when a tension is applied. The preparation of a homogeneous concentrated polymer solution is not easy. The particles of solid polymer can form a solvent-swollen congealed mass which restricts penetration by further solvent and is difficult to disperse. Mechanical mixers are used, and it may be necessary to heat the solvent and to exclude air if the polymer is sensitive to oxidation.

In practice, spinning solutions are in the range of about 15–30%, solids depending upon the polymer.

The ideal solvent for dry spinning should have a low boiling point, and low latent heat of evaporation. It should be readily available in pure form, and cheap. Chemically, the solvent should be inert, thermally stable, and non-toxic.

Pigments and other additives may be incorporated, and the solution then filtered and pumped to a storage tank. Here entrapped gas bubbles clear from the solution, and the blending of batches can take place. To reduce the viscosity and make pumping and filtration easier, the solutions are warmed.

Gear pumps meter the liquid through a final filter pack and to the spinneret. As for melt spinning, filter and spinneret are assembled in a common unit (spinning head), and are heated by circulating liquid or vapour. This final filter is needed to remove small particles— much as those present in melt spinning—which might block the jets or cause yarn defects.

The filaments are extruded vertically downwards into a heated tube 5 or 6 m long (the spinning chimney). Here they are exposed to a circulating current of hot gas which evaporates solvent from the surface. Hot air or inert gas (nitrogen or steam, for example) is used, but great care is needed when using air that the concentration of solvent vapour does not reach the explosion limit. There is a very real danger of fire or explosion in a plant with large quantities of flowing hot organic vapours, and high dielectric constant fibres running continuously over guides. Careful precautions are taken to avoid sparks.

Temperatures are chosen to match the physical properties of the solvent and to provide optimum spinning conditions. As illustrations, some published figures for spinning cellulose acetate from propanone (acetone) give a spinneret temperature of 56°C and chimney temperature about 95°C at gas inlet, and 60°C at outlet. For an acrylic fibre (Chapter 8) spinning from dimethylformamide (bp 153°C) corresponding temperatures might be about 130°C, and 250°C at the gas inlet.

At some point down the tube the hardened filaments can be made to converge by a guide, spinning finish applied, and then wound up. Speeds of about 500 m/min^{-1} (and up to about 1000 m/min^{-1}) are employed. The wind up imposes a drawing-down tension which acts on the still fluid portion of the filaments, much as described for melt spinning, but filament formation is more complex. (It should be made clear that the tension in the threadlines is mainly the result of air drag, the fluid zone could not support the tensions actually observed.) Solvent evaporates rapidly from the surface of the extruding liquid jet and causes concentration of the polymer at the surface to the limit of solubility. The polymer molecules then form an elastic jelly, swollen by solvent, which can be drawn down in diameter by the applied tension. Continued evaporation from the surface produces a hard skin of solid polymer, and effectively stops draw-down. The core of the filament is still liquid or jelly-like (gel). Diffusion of solvent out through the surface skin reduces the volume of the core and causes the skin to fold to give an irregular fibre cross-section, or even to collapse to a flattened 'dog bone' section. The degree of folding depends upon the relative rates of solvent evaporation and of diffusion to the surface. The slower the rate of evaporation, the more nearly circular is the transverse section of the filament.

Non-circular spinneret holes can be used, but to retain any imposed shape the skin must be formed very rapidly. The formation of the deformable gel state, before separation of a solid phase, is essential to the spinning process.

The most careful control has to be established over every minute detail of the process if good quality yarns are to be produced. Each filament needs the same constant conditions for solvent evaporation over each element of its path down the spinning chimney, just as in melt spinning uniform cooling conditions are necessary. The similarities between the two processes will be apparent, together with the essential difference and complexity of dry spinning in that mass transfer of solvent is involved.

The wound-up filaments retain small quantities of solvent, which may be washed out later. The degree of molecular orientation along the axis of the spun fibre is low. Much of the orientation which is introduced by the flow at extrusion and at draw-down has time to randomise in the gel state. Therefore, the filaments are given a hot stretching process, essentially similar to the drawing of melt spun fibres, but often to a greater extent. Stretching sets up internal stresses, and these may be released by an annealing heat treatment whilst holding the yarn to length.

If dry spinning is to be economic it is absolutely vital to have efficient recovery of the solvent. The vapour can be condensed out from the carrier gas, absorbed on carbon columns, or extracted by water or other liquid columns. Closed circuit spinning systems are used in which the gas circulates in a loop, one side of which is the spinning tube, and the other side a condenser system to recover solvent.

Dry spinning is used to manufacture continuous filament yarns, but it is not used for staple fibre which is more economically produced by the wet spinning technology. Dry spinning is slower than melt spinning, and solvent must be recovered, and it is not capable of the mass throughput of wet spinning.

WET SPINNING

Wet spinning begins with a concentrated polymer solution which is prepared and filtered according to the general principles considered above, and metered accurately to each spinneret assembly. The fluid jets emerge (usually horizontally) into a bath of a liquid which is not a solvent for the polymer, but which is miscible with the fluid in which the polymer is dissolved.

The surface of the issuing jet forms an interface between polymer solution and bath liquid. Solvent diffuses out from the surface, and the non-solvent bath liquid diffuses into it. A thin surface film of polymer gel is formed very rapidly; the process of gelation is known as coagulation. Continued exposure to non-solvent brings about phase separation of solid polymer as a surface skin. As the two-way diffusion proceeds, the gel layer is carried through to the centre of the fibre, followed by a thickening of the skin of the solid polymer.

The hardening filaments are being drawn away to be wound up, and the tension on the filaments (which are now supported by the bath liquid) produces draw down in diameter immediately upon emergence from the holes, and before they skin over. The basic pattern of spinning is repeated, superficially at least.

There are more opportunities for subtlety here. By dissolving chemical reagents in the bath liquid, reactions can be carried out within the nascent filaments. Viscose rayon is produced by such a combination of coagulation and chemical reaction, and the wet spinning process for rayon is described in more specific terms in Chapter 5.

In wet spinning, as in the other spinning technologies, the filaments must be given uniform treatment if good quality yarns are to be spun, and fresh supplies of bath liquid have to be provided in quantity. Obviously, this is a more economic proposition if water can be used, and aqueous systems are used for all commercial fibres. Spinning from an organic solvent straight into water can cause too rapid precipitation of solid polymer, and commonly the coagulation bath contains an aqueous solution of the organic solvent to delay precipitation. Similar procedures may be followed when spinning polymer solutions in concentrated acids, the coagulation bath may be a dilute acid. Inorganic salts are also frequently added to promote and control coagulation and solidification. Following the coagulation and hardening bath, there may be further baths to complete the process of hardening and washing out solvent. The actual path length of the filaments in the spinning bath can be quite short—less than one metre—then they are hard enough to be handled on rollers, and subsequent washing and hardening can be carried out on rollers.

The temperature of the coagulation bath can also be varied, to modify the spinning process.

The bath liquid must be provided without turbulent flow so that the extruding filaments are steady and do not break or touch each other until they are sufficiently hardened.

The filaments are oriented by stretching, which may be carried out in a bath, but which is usually done as soon as the filaments leave the spinning bath by the method of stretching between two rollers rotating at different speeds. In this state the filaments are swollen by solvent, and non-solvent bath liquid, which plasticise the polymer molecules—increase their mobility—and facilitate orientation. Alternatively, or additionally, the filaments can be hot stretched.

The spinning speeds are restricted by the considerable viscous drag on the filaments in the bath liquid, and by the time needed for coagulation and hardening. However, whilst there must be access of bath liquid to all filaments, the process of filament formation is not restricted by heat transfer. Thus, very large numbers of filaments can be extruded from one spinneret plate—of the order of ten thousand—and used to make staple fibre. Practical spinning speeds may be no more than 50–100 m/min^{-1}, but the number of filaments produced can result in high productivity. (For high quality continuous filament yarn the number of spinneret holes would be no greater than 100, so that each filament experienced uniform conditions and was uniform in properties.) The spinneret plates do not experience very high pressures and are of thinner metal, but may have to withstand corrosive liquids and so are made of expensive platinum or tantalum alloys. The hole diameters are of the order of 10 μm— one tenth the diameter of the melt spinning hole diameters. Solvents and chemicals are recovered from the spinning bath and wash water.

Further consideration of the wet spinning process and of the technology is given in Chapter 5.

BICOMPONENT FIBRES

The above discussions have considered the extrusion of a single polymer component melt or solution. Mixtures of polymers can be spun also, but as yet have little application. However, means have been devised for extruding two distinct polymer streams through the same spinneret hole. Filaments can be made with the two components side-by-side, or with one polymer down the middle as a core to a sheath of the second. Both solution and melt spun bicomponent fibres are made in this way.

There are several ingenious devices for bringing together the two polymer feeds in the correct manner, but the principle of the process is not as difficult as might at first appear, for in the absence of agitation there is little tendency for the viscous polymeric liquids to mix. A sharp interface can be maintained. Perhaps a greater problem is to make the two stick together when the fibre is spun and drawn.

Some uses for these unusual fibres will be described in Chapter 12.

4

The Production and Preparation of Natural Textile Fibres

COTTON

The use of cotton fibres in the ancient world was widespread in India and other Asian countries, in the Middle East, and in South America. The earliest archaeological record of a cotton fabric is from about 3000 B.C. in India, and the cultivation of cotton was practised in India certainly by 1000 B.C. Cotton had little impact on the English textile scene, however, until the mechanical inventions of the eighteenth century and the growing of cotton in the American Colonies allowed the great expansion of the cotton industry, which flourished especially in Lancashire, and which was the spearhead of the Industrial Revolution. The invention of the flying shuttle by Kay (1738), the spinning jenny (Hargreaves, 1764), the spinning frame of Arkwright (1769), the mule of Crompton (1779), and the cotton gin of Whitney in 1794, powered eventually by the steam engine, permitted the mechanisation of the textile industry. In the nineteenth century cotton displaced wool as the major textile fibre; it holds this position still, with an annual textile usage of the order of 10^{10} kg, providing more than 50% of all textile fibres used in the world.

Cotton is the seed hair of the plant, growing inside the seed pod, or boll. The hair is a single-cell cylindrical filament which collapses to a flattened and twisted fibre, as the boll opens and dries (*Figure 4.1*). As well as the long hairs, short, coloured hairs called 'linters' are formed. The cotton seed is picked by hand or machine (the latter especially in the U.S.A.). About 120–150 bolls yield 1 kg. The seed is mechanically processed to separate the hairs from the seed, linters, and other vegetable debris and dirt—a process known as 'ginning'. The seeds are a source of edible oils and cattle feed; the linters provide cellulose fibre for paper, rayon, and other chemical manufacture.

The separated cotton fibres are baled for transport. Textile processing begins by blending several bales to even out quality, followed by further mechanical cleaning. The length of the fibres depends upon type and quality, but ranges from about 1.2 cm ($\frac{1}{2}$ in) for the short Asiatic cottons to about 5 cm (2 in) for the longest Sea Island types. Most cotton used is American (*Figure 4.1*), about 2.5 cm staple length; the main supply of longer staple is Egyptian cotton (about 3.8 cm). In general, the longer the staple the finer it is. Long staple also allows more overlap of fibres in the yarn and so can be

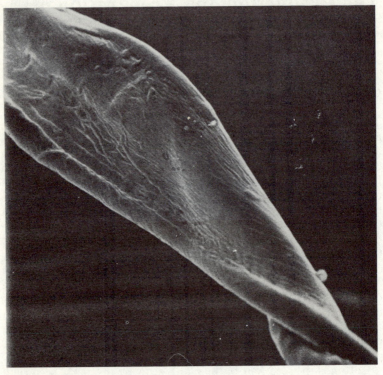

Figure 4.1   Stereoscan photograph of an American cotton fibre, magnification × 2960

spun to stronger yarns, or to finer yarns at a given strength. The convolutions of the flattened hairs provide inter-fibre cohesion and facilitate spinning.

The fineness of a spun cotton yarn is measured by the cotton count (British) which defines the number of hanks of 840 yards in 1 lb of yarn—this is converted to tex by dividing the count into 590.5 (we shall make no use of 'count').

It should be noted that no washing or chemical treatment is necessary before cotton is spun, the cotton hair being almost pure cellulose. The chemical treatments follow after making up into fabric. Fabrics are scoured (washing is always referred to as scouring) in either an alkaline soap, or a dilute sodium hydroxide solution, which removes natural wax and adventitious grease and dirt. Cotton warp yarns for weaving are treated with size—a starch-based solution containing a lubricant—which functions as a protective, lubricating, and stiffening agent during processing, and sticks down loose fibres. Warp sizing is a general practice, not confined to cotton yarns. It is usual to remove the size from cotton fabrics before scouring either by careful hydrolysis of the starch with dilute acid, or with enzyme preparations.

Cotton is bleached with either sodium chlorate(I) (hypochlorite) solution, prepared by passing chlorine gas into sodium hydroxide solution, or by hydrogen or sodium peroxides.

Some fabrics are mercerised, a treatment named after its inventor, J. Mercer (1850), which involves immersion in 20–25% sodium hydroxide solution at 30–40°C for 1–3 min. The fibres swell and untwist, returning to a more cylindrical shape, with shrinkage in length. The strength and lustre are improved and the fibres take up more dye giving deeper and brighter shades.

Various resins and other treatments may be applied as desired to impart crease resistance, water-proofing, and other properties.

As a textile fibre, cotton has the advantage of low price and ready availability, and can be efficiently processed to yarns and fabrics with a great variety of structures and uses. The fibre is lacking in resilience, and properties of handle and drape are not very good, but cotton garments have excellent water transport properties and are very comfortable to wear, even under humid conditions. Cotton is easily dyed with a wide range of dyestuffs.

OTHER VEGETABLE (CELLULOSIC) FIBRES

Very large numbers of other plant species have been used as a source of fibre. The seed hairs, leaf, stem and bark, and even fruit (coconut) of many plants provide useful fibres. As may be expected, use of plant fibres also goes back to pre-history and is common to primitive peoples throughout the world. The chief use of such plant fibres is for ropes, and for twine. Only the very briefest outline of the more important ones is given here.

Linen, from flax, was the principal European vegetable fibre until the rise of cotton in the eighteenth century. Its use is ancient, the early Egyptians were especially skilled in weaving fine linen cloths, and used it for mummy wrappings. The word 'line' derives from the same root, which perhaps stresses its influence on our civilisation.

The production of flax fibres is fairly typical of the methods used to separate other plant fibres. The stems are cut and soaked in ponds or open tanks for several days to bring about degradation of the non-fibrous substances by rotting (termed 'retting'). After drying, the stems are crushed and beaten ('scutched') to break the woody core and liberate the fibrous strands. The long fibre (30–90 cm) is combed and spun to provide yarn for fine linen fabrics. The shorter fibre is used for spinning coarser yarns, and for cordage. Approximately 700 000 000 kg of flax fibre is used each year.

Jute, ramie, and hemp fibres are obtained by similar processes. Jute is the most important of these vegetable fibres, and is next to cotton in terms of volume and diversity of use. The main area of cultivation is India and Bangladesh; the fibre is cheap, and used *inter alia* for sacking, webbing, industrial cloths, backings for rugs and carpets, and for twine.

Kapok is a seed hair which is buoyant, and impermeable to water —hence its use in stuffing life-belts, etc. The main use of kapok is for stuffing upholstery. The coconut fibre, called coir, is produced in Ceylon and used for bristles and door mats. Brush fibres are obtained from the stems of various palms.

WOOL

The term wool is applied to fibres from the fleece of the sheep. Other animal hairs are used in textiles, for example, of the camel, alpaca, and llama, of the Cashmere and Angora (Mohair) goats, and of the Angora rabbit. These are in limited supply, and very expensive. Wool is by far the most important animal fibre.

Wool and other animal fibres must also have been used by early man throughout the world. It has been the most important fibre in Europe, and was definitely in use in fabrics by about 1000 B.C. The wool trade became especially important to the prosperity of England, initially by the export of raw wool, but after about 1300 the domestic woollen industry became better established so that by the reign of Elizabeth I about 80% of our exports were woollen goods. Mechanisation, and growth of the modern industry sprang from the inventions of spinning and weaving machines, as listed above for cotton.

Wool protein is classified as a keratin, a group of fibrous proteins found in animal hairs, horns, nails, and quills. Because of its technical importance, a great deal of excellent scientific work has gone into the study of the very complex chemistry, physics, and biology of wool, and their interaction with textile properties and processing. A good contribution to this work has come from research workers in this country.

The fine wool fibre is not homogeneous in structure. As mentioned earlier, it is made up of a core of elongated cortical cells surrounded by a sheath of flat scale cells. (*Figure 2.1* shows the surface of a wool fibre taken with the scanning electron microscope.) The crimp of the wool fibre is decreased by wetting, and recovered on drying: these changes are caused by a differential swelling which seems to arise from side to side differences in properties across the fibre. The protein is also heterogeneous in chemical composition and in structure. The overall composition can be determined by hydrolytic breakdown of the protein into its constituent amino acids, followed by their separation and analysis. Prior to the development of the techniques of chromatography, this was a very lengthy and difficult task.

Eighteen amino acids are present. The most abundant of these are glycine [aminoacetic acid, $CH_2(NH_2)COOH$], serine [$HOCH_2CH(NH_2)COOH$], glutamic acid [$HOOC(CH_2)_2CH(NH_2)COOH$], leucine [$(CH_3)_2CHCH_2CH(NH_2)COOH$], arginine [$NH_2C(NH)NH(CH_2)_3CH(NH_2)COOH$], and cystine:

$$
\begin{array}{llll}
NH_2 & & & NH_2 \\
| & & & | \\
CH-CH_2-S-S-CH_2-CH \\
| & & & | \\
COOH & & & COOH
\end{array}
$$

Not a great deal is known of the precise order in which the eighteen amino acids are assembled in the protein molecules. The polypeptide structure of proteins was deduced by Emil Fischer at the turn of the century. The large molecules are built up from amino acids linked together by peptide bonds:

$$
\begin{array}{cccc}
R & R' & R'' & R''' \\
| & | & | & | \\
-NH-CH-CO-NH-CH-CO-NH-CH-CO-NH-CH-CO-
\end{array}
$$

here, R, R', etc., represent the groups characteristic of different amino acids). You will find that no synthetic fibre described in this book has the above peptide structure—the closest to it chemically are the polyamides of structure:

$$-NH-(CH_2)_x-CO-NH-(CH_2)_x-CO-NH-(CH_2)_x-CO-$$

The synthesis and study of polypeptides is of considerable significance to the development of polymer and biological science, and several have been spun from solution to give fibres with excellent properties. But, so far they are too expensive to make for commercial exploitation.

The acidic and basic amino acids in wool, for example, glutamic acid and arginine, introduce free reactive groups which can interact with dyestuffs. Probably they form internal ionic links with each other, within and between the protein molecules. Stronger, covalent bonds also bind the molecules together. These cross-linking bonds are disulphide from the amino acid cystine:

$$
\begin{array}{ccc}
R'' & & R \\
| & & | \\
-NH-CH-CO-NH-CH-CO-NH-CH-CO- \\
& | \\
& CH_2 \\
& | \\
& S \\
& | \\
& S \\
& | \\
& CH_2 \\
& | \\
-NH-CH-CO-NH-CH-CO-NH-CH-CO- \\
| & & | \\
R & & R'
\end{array}
$$

Thus, some protein molecules are not individual entities, but are bound together by covalent bonds into a three-dimensional network. These cross-links play an important role in the chemistry and physics of wool fibres. They can be opened by a variety of reagents, and then reclosed at a new position to set the fibre in a different configuration. Based upon this property, processes are in use to set wool fabrics to impart improved dimensional stability, to set in pleats and creases, and to bring about the 'permanent' waving of human hair.

The crystalline structure of wool was the subject of pioneering X-ray diffraction work by W. T. Astbury in the 1930s. Two crystalline forms were recognised. Unstretched fibre has a structure designated alpha, and this converts to the β-structure on stretching. Astbury interpreted the α-structure as a folded molecular conformation, and the β-structure as the fully extended molecules which are hydrogen bonded together along their lengths in sheet-like layers. Now the α-structure is known to be a helix, basically as proposed by Linus Pauling and R. B. Corey in 1951.

The largest producer of raw wool today is Australia, selling wool mainly from the Merino sheep which have been bred for fine, highest quality wool. Merino wool staple length is normally 2.5–10 cm and in the range of 12–25 μm diameter (wool fibres are not perfectly circular in cross-section). It has the highest number of crimps per cm (5–10), and the finest surface scale structure. The longer wool breeds of sheep (providing both meat and wool) produce a thicker and less soft wool. Staple length may be in the range 7.5–20 cm, and the wool is stronger and more resilient. The longest staple (15–40 cm) is generally the most coarse, the strongest, most resilient, and of good lustre. Mountain and hill breeds (for example the Scottish Blackface) provide this wool which is used mainly for carpets. In contrast to the Merino, a coarse wool may be 30–60 μm in diameter, with about 1 crimp per cm, and a coarser scale structure.

A very wide range of types and qualities of wools thus arises from the many different breeds of sheep, within the breed, and from different parts of the same fleece. The grading and sorting (by hand) of wool according to quality (length, fineness, lustre) is a highly skilled task. Quality determines the end-use to which the raw wool is directed.

Although wool is widely available, it is more difficult and more costly to produce than cotton and is available in smaller quantities (the best Australian yields are only about 5 kg per animal). Thus, it is expensive. Because of its cost, woollen rags are recovered and shredded, and waste is recovered from wool combing processes to provide short staple fibre which may be blended with new wool.

The fleece from the sheep may contain as much as 60% by weight of wool wax, suint, and dirt and entangled vegetable matter. Wool wax and suint are derived from natural secretions from the skin of the animal; the wax is soluble in ether, the suint is mainly water-soluble potassium salts of fatty acids. Scouring in warm detergent solutions removes both; alternatively, extraction with organic solvents (e.g. trichlorethene) is used to remove wax, followed by a water wash. Solvent extraction has the advantage of causing less entanglement and felting (brought about by mechanical agitation in the aqueous scouring processes), and gives better yields of wool grease. The wool grease is recovered from both aqueous and solvent processes and provides a source of materials with quite extensive applications in products mainly in pharmaceuticals, cosmetics, and lubricants.

Excessive vegetable debris not removed by scouring is destroyed by 'carbonising'. The wool is wetted with 8% sulphuric acid solution and then heated in air at 80–90°C. Cellulosic matter is hydrolysed and degraded to a black, friable mass which is easily crushed and shaken out. (Wool is undamaged by this seemingly drastic treatment.) After heating, the wool is washed with water and then in dilute sodium carbonate to neutralise any free acid. The carbonisation process may be applied to either raw wool or fabric.

Clean natural wool has a yellowish tinge, so when required for whites or dyeing to pastel shades it must be bleached. In practice most woollen goods are coloured, so only a small proportion of woollens are bleached. Bleaching of any textile, of natural or man-made fibre, requires care as the reagents used are capable of causing chemical damage, and thus each fibre has its preferred bleaching conditions and reagents. Wool can be bleached at any stage—fibre, yarn, or fabric—and hydrogen peroxide is normally used. The general avoidance of alkaline solutions for treating wool should be noted, in contrast to cotton processing where acids are avoided. Alkaline solutions can open the disulphide cross-links of wool, and hot alkalis may even dissolve it.

Woollen fabrics require a further scour after manufacture to remove added processing lubricants, and this would be carried out before bleaching. Chemical treatments to impart crease resistance, moth proofing, etc., may be given to the fabrics.

The spinning of wool fibres to yarn is carried out on two basic systems—the woollen and the worsted systems. Essentially, the woollen system spins the shorter staple wools and provides yarns for woven fabrics such as tweeds, blankets, and some knitwear. The worsted systems process the longer staple and provide the more expensive finer yarns for woven suitings, knitwear, and hand knitting. The fibres are combed before spinning on the worsted system so that they lie parallel to the yarn direction. The woollen system yarns are more hairy, a result of the more random array of fibres in a freer structure, and are softer and lighter. Further description of yarn spinning processes is deferred until Chapter 12.

The 'felting' properties of wool are unique, and are exploited as a fabric finishing process to give wool fabrics a different texture, body, and appearance, as well as to make felts. The mechanism of felting is complex, but is a result of the elasticity of wool and of the scale structure of the surface. The scales project from the surface towards the tip of the fibre, and introduce frictional differences in motion towards the tip or towards the root. A wool fibre (or a hair) rubbed between finger and thumb along its length will travel in the direction of its root. The felting process applied to fabrics is known as milling, or fulling, and involves the intermittent mechanical compression of the fabric using rollers to feed it continuously through a restricted compression zone, or using hammers—a relic of the old process of treading the cloth underfoot. Milling is carried out usually in warm soapy water. The mechanical action brings about fibre migration, the elastic recovery from compression and the ratchet effect of the scales brings about consolidation of the fibres.

Unwanted felting and shrinkage of woollens can occur during

domestic laundering and several chemical treatments to reduce felting have been devised. These involve attacking the edges of the scales, for example with chlorine, or covering them by a resin treatment. Shrinkage of wool, as for all fibres, is a consequence of the relaxation of strains imposed during processing. Wool and cotton cannot be heat-set like the synthetic fibres, but woollen fabrics can be relaxed by controlled shrinkage. (Shrinkage brought about by mechanical felting is of course unique to wool because of its scales.)

World wool production is about 17×10^8 kg per annum at present.

SILK

Silk was the fibre of ancient China, its use and production gradually spreading through Asia to the Mediterranean and to Europe. Today silk is produced mainly in China and Japan at a rate of about 4×10^7 kg per year. It is extruded as the cocoon fibre of the larva of the *Bombyx mori* moth. The larva after feeding on mulberry leaves for about four weeks, enters the pupal stage and spins the silk about its body. The larva has two silk glands feeding one spinneret, and so produces twin continuous filaments which are embedded in a substance called sericin. As well as sticking the two filaments together, the sericin bonds together the cocoon structure. Both sericin and the silk fibre (fibroin) are protein in character. The fibre is extruded as a liquid which hardens on contact with air and as it is pulled and stretched by the moving head of the larva. A beautiful lesson from Nature on how to spin a two-component fibre and make a bonded structure in one step (consideration of such man-made fabrics is given in Chapter 12). In addition the twin filaments are roughly triangular in cross-section—note Chapter 3. They can be separated by dissolving the sericin gum in hot soapy water.

The pupae are killed by heat or steam and the cocoons are placed in hot water to soften the sericin. The silk threads are then brushed from the surface until a layer of good quality silk is reached and this is then wound off into skeins on a reel, combining the silk from several cocoons to maintain a uniform thread. The reeling process requires considerable skill. Each cocoon may contain 3 km or more of silk, of which from 500 to 1200 m may be reelable. The waste silk and damaged fibres are combed and spun as staple fibre, like cotton.

Sericin is removed either from yarn or finished fabric by a scour in hot, mildly alkaline (pH below 11) soap solution, the full beautiful lustre and softness of the silk fibre is then revealed.

Silk fibroin is a less complex protein than keratin, and the fibre has a uniform, non-cellular structure. It is composed mainly of three amino acids: glycine, serine, and alanine [aminopropanoic acid, $CH_3CH(NH_2)COOH$], with minor quantities of about fourteen others. The three principal amino acids have small side groups, and long sequences of them probably make up the crystalline regions of the fibre, which have the β-structure of extended molecules, hydrogen bonded together. The tenacity and modulus of silk are, therefore, greater than typical values for wool (which has the helical α-structure in the unstretched state), but the extensibility is less. The fibre is about 10 μm in diameter. If silk is dissolved in lithium bromide solution and reprecipitated, it is converted to the α-structure, and is then soluble in water—there are no covalent cross-links as in wool. A synthetic fibre which demonstrates the profound effect of crystallinity upon solubility is described in Chapter 8.

The high cost of silk restricts its use to the more expensive fabrics, to which it is suited, for it has excellent properties of strength (3–5 g dtex^{-1}) and of drape, combined with a warm, soft, smooth handle. The so-called 'scroop' of silk, the characteristic rustle of silk fabrics, is not a natural property, but is obtained by treating the fabric with dilute organic acids, for example acetic acid.

5

Cellulosic Fibres

Cellulose is the most abundant organic chemical in Nature, and vast quantities are produced each year by photosynthesis, but not all forms of vegetable life can be used as suitable or economic sources of cellulose as raw material for fibres. Conifers, which can be grown in Northern forests on land unsuitable for agriculture, provide the main supply. Cotton linters, fibres too short for direct spinning, offer a supply of purer cellulose, but this is offset by a lack of stability in price and availability, and wood has become the principal source.

Some of the reasons for the attraction of man-made fibres have been given in Chapter 1. Whilst natural cellulose fibres will continue to be spun in enormous quantities from cotton hairs, it is expected that the future pressures of civilisation for land and for food will limit the expansion of the growth of cotton, so we may expect regenerated cellulose fibre production to continue to grow in the future.

Up to 50% of wood may be cellulose, with about 25% each of lignin and of carbohydrate other than cellulose. These latter materials are dissolved out. Treatment of wood chips with calcium hydrogen sulphite (calcium hydroxide plus sulphur dioxide) and pressure cooking in steam for 12 h or so, is a commonly used process. Lignin, which is a complex polyphenolic resin, is partially degraded and solubilised. The resulting aqueous pulp of short cellulose fibres is washed, concentrated to about 30% cellulose, bleached, and then converted to a paper board for transport. A better quality cellulose is obtained by pretreatment of the wood chips with acid, followed by heating with alkali.

The essential chemical step for making fibres is then to bring the cellulose into solution so that it may be extruded. As mentioned in Chapter 2, two methods are in large scale use today, the viscose rayon and the acetate processes. There have been others, but these two have survived commercially, and we shall confine our attention to them.

VISCOSE RAYON

Courtaulds bought the British Patent rights and pioneered the technical and commercial development of Cross and Bevan's process; the first viscose rayon plant was built at Coventry in 1905 and the modern man-made fibre industry of the world was born.

Today world production of rayon is about 35×10^8 kg, almost half the total of all man-made fibres, but still quite small compared with the 10^{10} kg of natural cotton fibres which are processed by the textile industry.

Naturally, since its introduction many modifications and improvements have been made, and will continue to be made, to the viscose process. But, the essentials of the process are first to treat the wood pulp paper board with sodium hydroxide solution for perhaps $\frac{1}{2}$–1 h to swell the cellulose fibres, and then shred the board to fluffy 'crumbs'. The shredding stage is necessary to permit more uniform penetration and attack of the cellulose fibres in subsequent chemical treatment. The crumbs are 'aged' by holding for a few hours at a controlled temperature during which oxidation of the cellulose lowers the polymer relative molecular mass. The time of ageing may be shortened by catalysing the oxidation with cobalt(II) or manganese(II) ions. Careful control of this step is especially important as it is a degradation process. Treatment with carbon

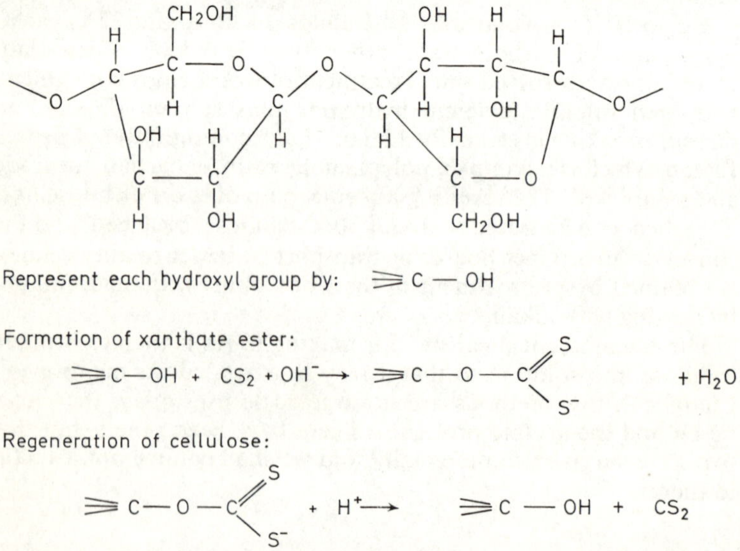

Represent each hydroxyl group by: $\equiv$ C — OH

Formation of xanthate ester:

$$\equiv C-OH + CS_2 + OH^- \longrightarrow \equiv C-O-C{\overset{S}{\underset{S^-}{\diagup}}} + H_2O$$

Regeneration of cellulose:

$$\equiv C-O-C{\overset{S}{\underset{S^-}{\diagup}}} + H^+ \longrightarrow \equiv C-OH + CS_2$$

Chart 1   Viscose process

disulphide then converts the alkaline cellulose crumbs to sodium cellulose xanthate (orange-yellow in colour), which is soluble in sodium hydroxide solution. This viscous solution is called 'viscose'.

The degree of conversion to xanthate of the three hydroxyl groups per glucose repeat unit of the cellulose molecule is low, and depends upon the quality of fibre to which the viscose is to be spun (see below). Cheaper ordinary grade rayon is spun from xanthate levels of about 0.5 groups per unit: high tenacity rayons from compositions of 1 to 1.5 xanthate groups per unit.

The chemical changes which occur are outlined in Chart 1, but the fine detail of the actual process, involving heterogeneous reaction of solid cellulose fibres with liquid reagents, is complex and probably incompletely understood.

Several batches of viscose are blended together to even out batch variations arising from cellulose supply and treatment, and the solution is then 'ripened' by holding at constant temperature (10–18°C) for up to 5 days. During this stage a more uniform distribution of xanthate groups among the cellulose molecules occurs. Finally, before spinning the solution is filtered and, if desired, pigments are added. It is then de-aerated to remove bubbles which would give flaws in the filaments.

Extrusion takes place into dilute (10%) sulphuric acid which restores the sodium cellulose xanthate to cellulose. As the viscose emerges through the spinneret holes into the acid bath the surface of contact is at once coagulated; regeneration of cellulose by chemical reaction with acid follows at a slower rate to build up a surface skin. A complex interaction of diffusion of ions into and out of the nascent filament is set up, through the coagulated gel membrane. As the alkaline viscose is neutralised the coagulated layer passes through to the centre of the filament, it is followed by regeneration of cellulose and formation of the molecular fine structure of the solid fibre. Contraction within the surface skin of regenerated cellulose produces a wrinkled fibre surface, which gives good covering in fabric, and greater dyeability. Whilst in this coagulating gel stage the filaments can be stretched to increase the orientation of the cellulose molecules along the fibre axis and so give improved physical properties.

Because of the complexity of the viscose process, it has offered a great deal of scope for changes to produce a variety of products. The developments in rayon technology have come through a better understanding and control of the detail of the process, and by the use of chemical agents either in the viscose or in the spinning bath which modify the intricate physics and chemistry of the regeneration of cellulose fibres.

The spinning bath of dilute sulphuric acid contains sodium sulphate to bring about coagulation, and a small quantity of zinc sulphate. The mode of action of the zinc is not perhaps completely established, but it probably forms an insoluble zinc cellulose xanthate salt and retards the regeneration of cellulose. There is a longer time in the gel state during which the fibre can be more highly stretched. The use of zinc has been known since 1911. In the 1950s other components were introduced to control the rate of regeneration of cellulose. Known as modifiers, these include certain amines and poly(ethane diol), and are usually added to the viscose. Methanal (formaldehyde) may also be added to the spinning bath as a retarder and cross-linking agent. A high level of xanthate in the viscose also slows down the processes of regeneration.

The use of modifiers has enabled greater control of the crystallinity and fine molecular structure of the rayon fibre, and allowed higher degrees of orientation by greater stretching in the gel state, and therefore better physical properties. The filaments are more uniform in structure and properties through their cross-section, and are more nearly circular in section.

Additional improvements came from the use of purer cellulose pulp, obtained with less degradation of the high relative molecular mass natural cellulose, and of course from developments in the mechanics of spinning. Early rayon fibres had a maximum strength of about 1 $g/dtex^{-1}$, whereas today more than 7 $g/dtex^{-1}$ can be attained. These developments enabled rayon to capture the huge automobile tyre cord market from cotton.

It is interesting to note that in the 1950s and 1960s rayon tyre cords in turn began to give way to nylon, as the demand for tougher cords continued. However, more recently with the emphasis switching to radial-ply tyre construction for cars, and away from the conventional bias-ply tyre, rayon has recovered some ground. The radial-ply tyre has a stiff 'breaker belt' of tyre cords running circumferentially, and a flexible side-wall. The necessary stiffness of the breaker belt has demanded higher modulus cords. Rayon has a higher modulus than nylon. However, other higher modulus fibres—polyester, glass, steel, and new organic fibres are now jostling for future supremacy in this large market.

Having glanced at the chemistry, we must return to the technology of the spinning process. The filaments are pulled away from the spinneret through about half a metre of bath acid and can then be brought together at a guide and taken out and stretched between two godets which are rotating at different surface speeds (*Figure 5.1*). In the early processes, the stretched filaments were then wound up by centrifugal force on the inside of a rapidly rotating cylinder—the

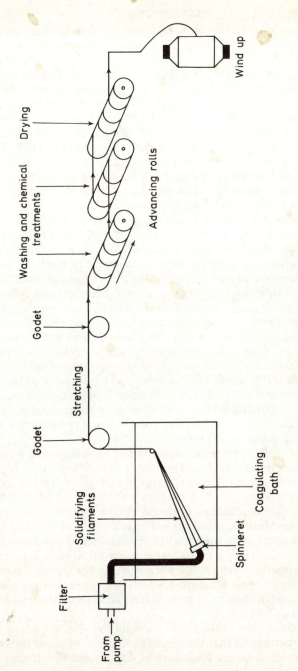

Figure 5.1 Diagrammatic representation of wet spinning

Topham box, after its inventor. The yarn was then further processed —neutralised, bleached, etc.—in batches. Modern processes are continuous. The stretched filaments are carried forward on advancing rolls. These are pairs of rollers with axes very slightly inclined together. The yarn runs backwards and forwards between the two, and is advanced along the rollers each time by the inclination of the axes. In this way, many hundreds of metres of continuously moving yarn can be accumulated for several minutes. Various wet treatments, and drying, can be carried out on each pair of a series of such rollers. (An even more compact arrangement is used, in which the two rollers are cut down to fingers and mounted so that they interlace with one another on almost coincident axes. The yarn then moves along as the two rotate together.)

The filaments are first washed to remove acid, then treated with sodium sulphide solution to remove sulphur residues of the xanthate, washed, bleached (if required), and dried. A finish is applied, and the yarn wound up on a package with the insertion of twist.

When modifiers are used to slow down the regeneration of cellulose, for example for tyre cord yarns, the filaments are especially vulnerable to breakage in the bath. Turbulence and hydrodynamic drag on the filaments can be reduced by spinning through a tube. The bath acid travels along with the filaments and higher spinning speeds are possible. The high tenacity yarns are stretched in hot dilute sulphuric acid, which completes regeneration of cellulose.

Rayon for textile staple fibre imposes much less severe restrictions on processing and on quality, and cheaper, lower relative molecular mass, lower xanthate, viscose can be spun using continuous processes in large capacity plant to produce cheap fibre. Typical staple spinnerets may contain 30 000 holes.

The economics of the viscose rayon process are helped by recovery of carbon disulphide, acid, and sodium sulphate from the spinning bath. Spinning speeds within the region $100-150 \text{ m/min}^{-1}$ can be achieved.

The attraction of cellulosic fibres lies in their low price and in their high moisture absorption, 13% at 65% relative humidity and 20°C, which gives comfort in wear next to the skin, ready processing and dyeing. However, viscose rayon staple is inferior to natural cotton in physical properties, especially when wet. In the 1950s improved cellulosic fibres were developed which are more 'cotton-like' in properties, and which have been given the distinguishing generic name 'polynosic fibres'.

Polynosic fibres are produced following the viscose process, but with modifications so that the relative molecular mass of the cellulose is maintained higher than normal. The viscose is extruded into a bath which is more dilute in sulphuric acid, regeneration and precipitation of cellulose are greatly slowed and ultimately give rise to a microfibrillar crystalline structure more analogous to that of natural cotton. The filaments are stretched out of the bath, and then pass to a second acid bath to complete the regeneration. The fibres are highly oriented, with an extension to break of about 10%, and their properties are much less sensitive to water. They are usually produced as staple fibre.

Rayon is an extremely versatile fibre and is widely used in textiles, especially blended with synthetic fibres to provide moisture absorption for greater comfort. Well over half of the viscose rayon produced is in the form of staple fibre. A wide range of deniers and several different fibre cross-sectional shapes—round, flat, multi-lobed, and serrated—are available in straight or crimped fibre. Uses include all types of clothing, domestic curtains, upholstery, covers, blankets, sheets, carpets, non-woven fabrics, disposables, paper, and medical and surgical applications; the high tenacity fibres have tyre cord use (especially for radials), and industrial uses for hoses, tapes, and belts, etc. In most of these uses involving staple fibre, the viscose may be blended with other fibres. Ordinary rayons suffer from poor physical properties when wet, and from shrinkage and loss of handle and garment shape. But fabric treatments can be applied to impart shrink and crease resistance. Polynosic fibres extend further the penetration of rayon into the traditional markets of cotton, and mark the possible future developments of improved cotton-like cellulosic fibres.

CELLULOSE ACETATE FIBRES

Natural cellulose may be brought into a soluble form by conversion to the acetate ester. Wood pulp is treated with acetic acid to swell the natural cellulose fibres, and the mass is reacted with acetic acid and acetic anhydride in the presence of sulphuric acid catalyst. The acetylation reaction is exothermic and the reaction vessel has to be cooled (below 50°C) to prevent excessive degradation of the cellulose. In one variation of the process dichloromethane is added with acetic anhydride to remove the heat of reaction by ebullition. (The bp of dichloromethane is 42°C.) The fibrous pulp dissolves to a clear viscous solution of cellulose triacetate (Chart 2).

Reaction is stopped by diluting with aqueous acetic acid. Some sulphate ester is also formed (which would subsequently liberate sulphuric acid in the fibre) and has to be removed by treatment with magnesium acetate. The solid cellulose triacetate may be precipitated by adding the solution to water; then it is washed and dried.

$$(Cellulose) - OH + (CH_3 CO)_2 O \longrightarrow (Cellulose) - O \cdot CO \cdot CH_3 + CH_3 COOH$$

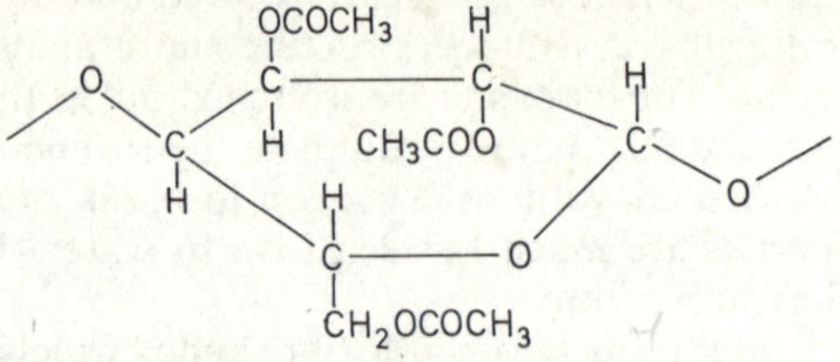

Chart 2   Formation of cellulose triacetate ester

Cellulose triacetate fibres are made by dry spinning a solution of the polymer in dichloromethane (plus a little methanol) into hot air.

When Cross and Bevan made the triacetate, they found that it was soluble in trichloromethane which was unsuitable as a spinning solvent. It was then discovered that if the triacetate is hydrolysed back to a composition about half-way between the triacetate and the diacetate it becomes soluble in propanone (acetone), a more readily available, cheap, and non-toxic solvent.

The cellulose acetate composition is referred to as 'secondary' acetate, and the fibre is dry spun from solution in propanone which contains a little water into hot air. The fibres are stronger, with a better handle, and a higher moisture uptake than the triacetate fibres.

Hydrolysis of the cellulose triacetate to the secondary acetate is accomplished by heating the reaction mixture after dilution with water at about 40°C for 10 h or so.

Secondary cellulose acetate was made in the U.K. for coating the fabric wings of first World War aircraft, by the Dreyfus brothers at Spondon. After the war, left with a large surplus capacity for cellulose acetate, they began to produce the fibre. The Company, British Celanese Ltd., is now part of the Courtaulds Group. Cellulose triacetate fibres were not produced commercially until about 20 years ago, when dichloromethane became available at an economic price.

Cellulose acetate will not dye with the dyestuffs used for cotton, because of the fewer hydroxyl groups and the danger of hydrolysing the acetyl ester groups in an acid or alkaline dyebath. As the first of the man-made fibres which was chemically different from a natural fibre, cellulose acetate presented a difficult problem in dyeing when it was first introduced. New dyes were developed for it, the disperse dyes. These are insoluble in water but can be applied dispersed in water. Chemically, they are mono-azo compounds or aminoanthraquinones. These disperse dyes were available for use with the synthetic fibres as they were later produced.

The cellulose acetate fibres are thermoplastic. The triacetate can be heat set and pleats can be set in it. In this respect it resembles the synthetic fibres nylon and polyester, but it has not the tenacity and toughness of these fibres.

Secondary acetate is spun in larger quantity than the triacetate, but still only about one tenth the scale of viscose production. Its good, soft handle has given it a wider market than triacetate in various clothing end-uses, mainly in dresswear. Triacetate is used for dress fabrics also, and has the advantage of drip-dry properties.

6

Polyamide Fibres

The world's first truly synthetic textile fibre was produced on pilot plant scale in 1938 by the Du Pont Company in the U.S.A. This fibre—6.6 nylon—was one result of research work begun in 1928 into the synthesis of polymeric materials, initiated and directed by W. H. Carother (Chapter 2).

Nylon is a generic name, not a registered trade name, covering a wide class of fibre-forming polyamide materials. The preparation of polyamides from a diamine and a diacid, or by self-condensation of an amino acid, has been mentioned in Chapter 2. The established large scale commercial nylon fibres are derived from linear aliphatic acids, diamines, and amino acids. Individual members are distinguished by recording the number of carbon atoms in the diamine component, followed by the number in the diacid component (including the carbonyl carbons), thus, the polyamide:

$$-[NH(CH_2)_6NH.CO(CH_2)_4CO]_n-$$

is designated 6.6 nylon, whilst the polyamide:

$$-[NH(CH_2)_6NH.CO(CH_2)_8CO]_n-$$

is 6.10 nylon. When the polymer is derived from a linear amino acid the single numeral defining the total number of carbon atoms is sufficient, thus, 6 nylon is the polyamide:

$$-[NH(CH_2)_5CO]_n-$$

By far the most important polyamide synthetic fibres are 6.6 and 6 nylons (also used on a much smaller scale as plastics). The reason for their pre-eminence is the ready availability of the six carbon atom compound benzene as cheap raw material for the synthesis of intermediates—together of course with superb fibre properties.

World production of all nylon fibres in 1969 totalled about 2×10^9 kg; by the end of 1971 production capacity was estimated at nearly 3×10^9 kg.

6.6 NYLON

The intermediates, hexane 1,6-dioic acid (adipic acid) and 1,6-diaminohexane (hexamethylene diamine), may be obtained starting from benzene. The original route passed via phenol by hydrogenation to cyclohexanol, which was then oxidised to hexane-1,6-dioic acid with 50% nitric acid (Chart 3). More direct modern routes to cyclohexanol (and cyclohexanone) are by catalytic air oxidation of cyclohexane, derived by hydrogenation of benzene. Cobalt salts are used as catalysts at about 150°C and at a pressure of about 0.9 MPa: yields are improved if oxidation is carried out in the presence of boric acid which is said to form borate esters of cyclohexanol and hinder continued oxidation. The further oxidation of the cyclohexanol/cyclohexanone mixture to hexanedioic acid is also carried out catalytically using air as oxygen source.

1,6-diaminohexane is obtained from hexane-1,6-dioic acid: the boron phosphate catalysed reaction of hexanedioic acid with

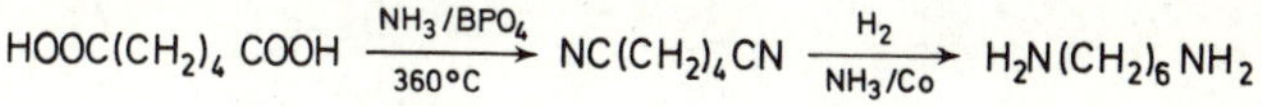

$$HOOC(CH_2)_4 \, COOH \xrightarrow[360°C]{NH_3/BPO_4} NC(CH_2)_4CN \xrightarrow[NH_3/Co]{H_2} H_2N(CH_2)_6NH_2$$

Chart 4   Diaminohexane from hexanedioic acid

ammonia at 360°C forms 1,4-dicyanobutane (adiponitrile) which is then hydrogenated to the diamine (Chart 4).

Other processes have been developed as routes to hexanedioic acid based upon furfural, CH—CH , obtained from oat husks

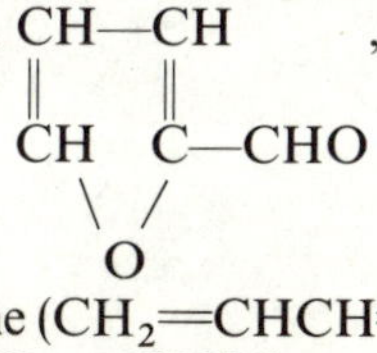

and straw, upon butadiene (CH_2=CHCH=CH_2), and upon cyanoethene (acrylonitrile, CH_2=CHCN) as the alternative starting materials to benzene. The impetus for the search for alternative routes came initially as the demand for benzene began to outstrip supply and to force up the price, but additional processes to obtain benzene from petroleum products have been developed to meet demand. The cost of making the intermediates for 6.6 nylon polymer is greater than for polyester (Chapter 7), and the stage of development of nylon is such that its cost structure is under pressure because of competition, although its use is still growing in volume. Thus, any real reduction in cost is especially important to the producer and routes to 1,4-dicyanobutane starting from cheap butadiene and cyanoethene are being fully explored, especially in the U.S.A.

The first Du Pont route from buta-1,3-diene followed the stages:

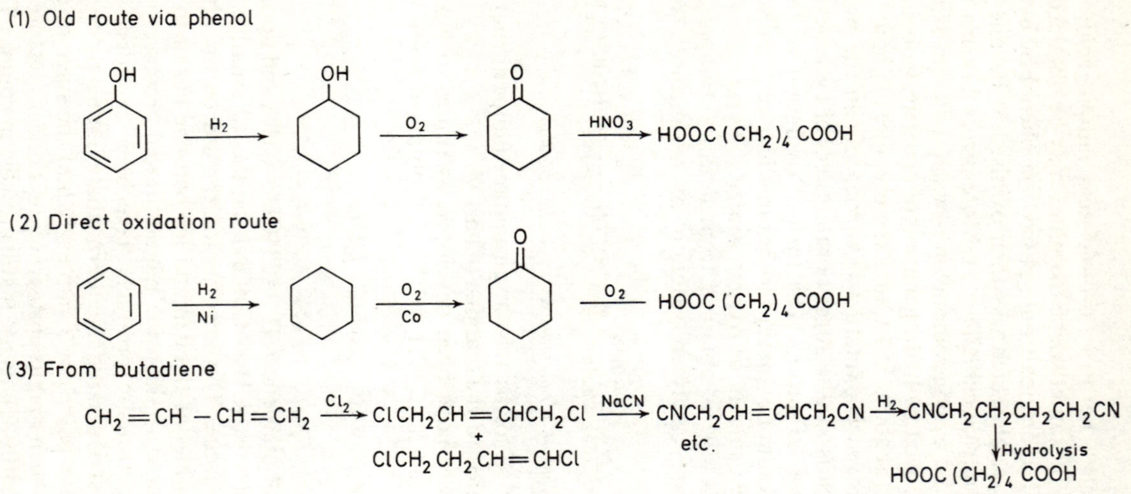
(1) Old route via phenol
OH
H2
OH
O2
O
HNO3
HOOC(CH2)4 COOH
(2) Direct oxidation route
H2
Ni
O2
Co
O
O2
HOOC(CH2)4 COOH
(3) From butadiene
CH2=CH−CH=CH2
Cl2
ClCH2CH=CHCH2Cl
+
ClCH2CH2CH=CHCl
NaCN
CNCH2CH=CHCH2CN
etc.
H2
CNCH2CH2CH2CH2CN
Hydrolysis
HOOC(CH2)4 COOH

Chart 3 Routes to hexanedioic acid

addition of chlorine to give 1,4-dichlorobutenes; reaction with sodium cyanide to give 1,4-dicyanobutenes; hydrogenation to 1,4-dicyanobutane (Chart 3). A more recent du Pont method is said to avoid the cost of the chlorine (which is not recovered) by a nickel-complex catalysed direct addition of hydrogen cyanide to butadiene to form dicyanobutenes. A similar route developed by Esso Research and Engineering involves reaction of butadiene with copper(I) cyanide and iodine to form a 1,4-dicyanobutene/copper/iodine complex, which is hydrolysed by aqueous hydrogen cyanide to liberate 1,4-dicyanobutene and regenerate copper(I) cyanide and iodine. Commercial use of these routes will presumably follow.

An interesting electrochemical process has been developed recently by the Monsanto Company. Cyanoethene, in an electrolyte solution, is reductively dimerised to dicyanobutane at a lead cathode:

$$2CH_2\!\!=\!\!CH.CN + 2H^+ + 2e^- \longrightarrow NC(CH_2)_4CN$$

If nylon polymer of fibre-forming quality is to be obtained, it is essential that the intermediates are of extremely high purity and are used in precisely equimolar proportions. A neat way of obtaining correctly balanced proportions (and giving an additional purification step) is to form the salt, $(NH_3(CH_2)_6NH_3)^{2+}(CO_2(CH_2)_4CO_2)^{2-}$, which crystallises from mixed solutions in methanol of the diamine and dioic acid.

Polymerisation of the nylon salt is carried out by a batch process in an autoclave. An aqueous solution of the salt is concentrated to about 85% in an evaporator, and then charged hot to the autoclave. Air is purged from the vessel and the temperature raised, and polymerisation begins. When the internal steam pressure reaches about 1.7 MPa (250 lbf in^{-2}) at about 220°C, steam is allowed to bleed off to maintain this pressure until a temperature of about 250°C is reached. The steam pressure is then slowly reduced to atmospheric, whilst the temperature continues to rise to a maximum of about 280°C. After a short holding time at atmospheric pressure, the molten polymer is extruded under nitrogen pressure from the autoclave through a slot, and quench cooled with water on a casting wheel to produce a broad ribbon of polymer. This ribbon is cut to fragments (chip) of size suitable for feeding to melt spinning machines.

The autoclave polymerisation cycles are controlled automatically and the whole process takes about 3 h. Batches of perhaps a tonne of polymer are made. The operation of the autoclave under pressure prevents loss of the volatile diaminohexane during the early stages, and the precipitation of solid polymer which would occur from a solution boiling at atmospheric pressure as soon as polymerisation started. The cycles are developed to keep the polymer molten at all times, and to obtain maximum productivity. (The melting point of dry 6.6 nylon is about 265°C.) Some loss of diamine from the autoclave will always occur, and has to be balanced by precisely controlled additions to the initial salt solution. Acetic acid to limit the relative molecular mass of the polymer, titanium(IV) oxide (delustrant), and other additives or pigments may be added either to the salt solution or to the autoclave, as required.

Delustring with titanium(IV) oxide is important to increase the whiteness and opacity of textile yarns and fabrics and to prevent undesirable glossiness. The usual range of addition is from about 0.1 to 2.0% by mass, depending on the fibre end-use, and a dispersed particle size of about 0.1 to 0.5 μm is desirable. Aggregation of titania particles can cause troubles at spinning, and the ionic nylon salt solution is especially likely to cause aggregation and therefore addition of titanium(IV) oxide is a critical step in manufacture.

Antioxidants may be added to protect some textile and industrial yarns from oxidation during subsequent textile processing involving exposure to hot air. Oxidation leads to a loss in strength and discoloration of the yarns, and affects dyeing behaviour by destroying amine end-groups (see below). Various organic antioxidants have been patented for textile yarns, whilst tyre cord yarn, which experiences the most severe conditions, is usually protected by incorporation of copper and halide salts, particularly iodides. Titanium(IV) oxide (in the Anatase crystalline form) promotes the photo-oxidation of nylon in sunlight but this is countered by adding manganese salts. Antistatic agents may also be added to the autoclave: these are usually poly(epoxyethane) derivatives.

The formation of polyamides is a reversible reaction which may be represented as:

$$-\text{COOH} + \text{NH}_2- \rightleftharpoons -\text{CONH}- + \text{H}_2\text{O}$$

The autoclave polymer is in equilibrium, or near equilibrium, with steam at atmospheric pressure and at the temperature of the autoclave. By spinning under a steam atmosphere this equilibrium can be maintained and provides an important means of fine control of the polymer amine and carboxyl end-groups, and hence of relative molecular mass. The careful control of the number of amine end-groups is most critical if yarns of uniform and consistent dyeing properties are to be made, for the basic amino groups are important sites for ionic interaction with acidic dyes. By changing the acid/base properties of nylon, yarns of widely different dyeing characteristics can be made. Deep dye and acid dye-resist yarns are marketed which, in appropriate combinations in a fabric, allow multicoloured patterns to be dyed from a single mixed dye bath. These extreme dye variants may be obtained by the copolymerisation of small quantities of suitably reactive basic or acidic substances with the 6.6 nylon.

The autoclave batch process for nylon polymer manufacture has some disadvantages, for example in the long heating time which introduces some side reactions and decomposition of the polymer, and initial capital outlay in pressure autoclaves is high. But it is economic because many different grades of polymer are required for a large variety of yarn end-uses. It also permits the blending of chip.

Continuous polymerisation, which may be followed directly by spinning the molten polymer without wasteful and harmful cooling, chipping, and remelting, is technically desirable and offers economic advantages for individual yarns with a very large volume market, i.e. the continuous polymeriser can be run continuously on one polymer grade. The enormous problems associated with the development of commercial continuous polymerisation to give high quality yarn direct from 6.6 salt have been overcome and some processes are being used. The factor limiting the time of polymerisation in the autoclave is removal of water from the melt, not the rate of the chemical reaction. An elegant high productivity I.C.I. process devised by J. A. Carter achieves high rates of polymerisation by pumping nylon salt solution through a specially designed, heated, stainless steel pipe; fully made, molten polymer emerges and may be spun directly.

6 NYLON

6 nylon also was made in the laboratory of W. H. Carothers, but it was developed to successful commercial production by P. Schlack in Germany, and spun in 1939 by I. G. Farbenindustrie. The outbreak of war led to the independent growth of the two polyamide fibres so that now, although world production of polyamide fibres is split roughly evenly between the two, 6.6 nylon has about 80% of the U.K. market and 75% of the U.S.A. market, whereas 6 nylon has about 75% of the Japanese market and is predominant in Europe. This pattern of production is, however, gradually disappearing.

The polymer is prepared industrially from hexanolactam (caprolactam), not from the amino acid, 6-aminohexanoic acid, $NH_2(CH_2)_5COOH$; the pure, dry lactam will not polymerise, but in the presence of a little water, and especially if either a little aminohexanoic acid or 6.6 salt is present also, polymerisation can be effected at 220–260°C.

$$n \quad \underset{\substack{CH_2 \\ | \\ CH_2 \diagdown \\ \diagup CH_2 \diagdown}}{\overset{CO-NH}{\diagup \diagdown}} \underset{\substack{CH_2 \\ | \\ CH_2}}{\diagdown} \xrightarrow{H_2O} \quad H\!\!-\!\!\left[NH(CH_2)_5\,CO\right]_n\!\!-\!\!OH$$

Because the reaction is reversible there is always some free lactam present in molten 6 nylon polymer. Whilst 6.6 and 6 nylons appear very similar when written as conventional chemical formulae, the involvement of lactam introduces marked differences in technology.

Polymerisation is believed to be initiated by hydrolysis of a little of the hexanolactam to aminohexanoic acid, which then reacts with lactam in a succession of addition steps at the amino group:

$$\underset{\substack{CH_2 \\ | \\ CH_2 \diagdown \\ \diagup CH_2}}{\overset{CO-NH}{\diagup \diagdown}} \underset{\substack{CH_2 \\ | \\ CH_2}}{\diagdown} + H_2O \longrightarrow NH_2(CH_2)_5COOH$$

$$\underset{\substack{CH_2 \\ | \\ CH_2 \diagdown \\ \diagup CH_2}}{\overset{CO-NH}{\diagup \diagdown}} \underset{\substack{CH_2 \\ | \\ CH_2}}{\diagdown} + NH_2(CH_2)_5\,COOH \longrightarrow NH_2(CH_2)_5CONH(CH_2)_5COOH$$

The reaction is acid catalysed, and the protonated lactam,

$$\underset{\substack{CH_2 \\ | \\ CH_2 \diagdown \\ \diagup CH_2}}{\overset{CO-NH_2^+}{\diagup \diagdown}} \underset{\substack{CH_2 \\ | \\ CH_2}}{\diagdown}$$

is probably the active species which reacts with the amino group.

Once polymerisation has been initiated normal condensation reactions of carboxyl with amino groups take place also, and, as in the case of 6.6 nylon, an equilibrium is set up between concentrations of end-groups and of water in the molten polymer.

Hexanolactam, like hexanedioic acid, is manufactured mainly from cyclohexane and of the order of 2×10^6 tonnes per year are produced throughout the world, for which the main outlet is 6 nylon polymer for fibres.

The general route is to oxidise cyclohexane to cyclohexanone (plus cyclohexanol which can be dehydrogenated catalytically to cyclohexanone), as for hexanedioic acid production. Reaction of cyclohexanone with hydroxylamine gives the oxime, which under-

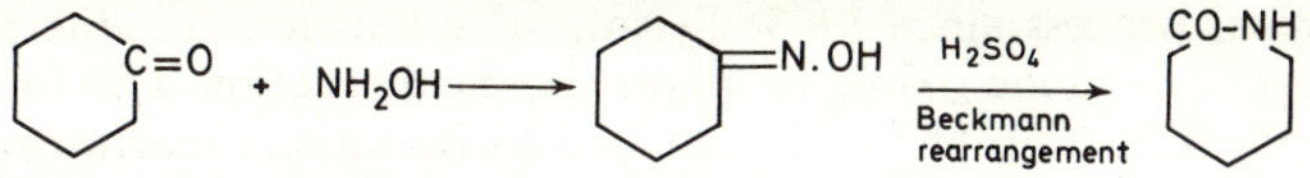

Chart 5   Hexanolactam from cyclohexanone

goes Beckmann rearrangement in solution in oleum to give hexano-lactam. These steps are summarised in Chart 5.

The Beckmann rearrangement has the following mechanism:

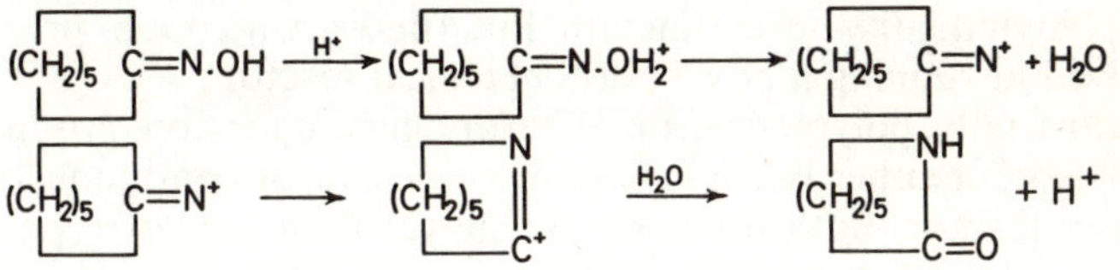

It is believed the oxime is protonated in the concentrated acid solvent, and then forms the positive nitrogen ion intermediate which rearranges to the carbenium ion structure.

To separate the hexanolactam at the final stage the sulphuric acid is neutralised with ammonia: ammonium sulphate may be formed also during oxime preparation, as hydroxylamine is usually prepared as the sulphate salt. The conventional route to hexanolactam yields as much as 4.75 tonnes of ammonium sulphate per tonne of lactam! The economics of the process are therefore greatly dependent upon the ability to sell ammonium sulphate at a reasonable price (for fertiliser).

(1) Techni-chem process:

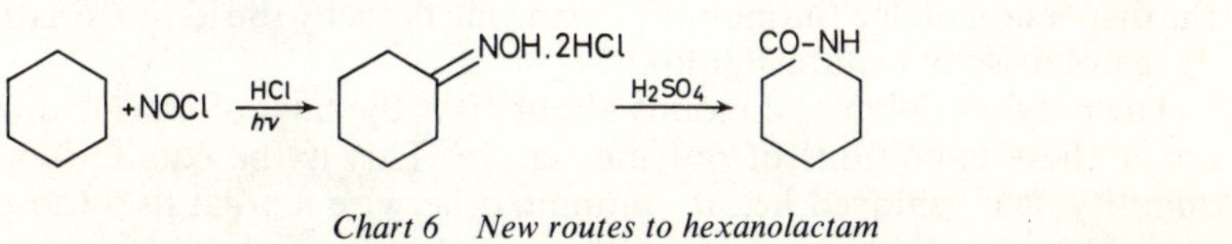

(2) Toyo photochemical process:

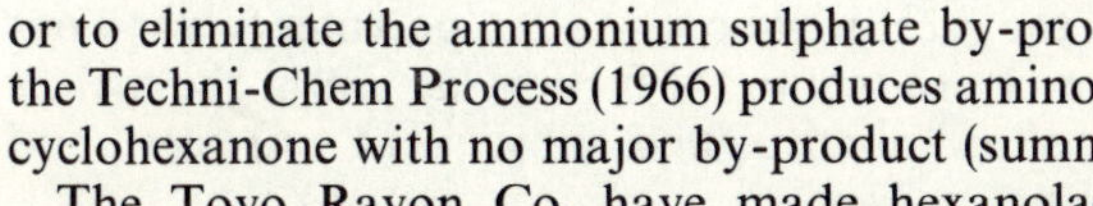

Chart 6   New routes to hexanolactam

Much effort has been directed at new or modified routes to cut down or to eliminate the ammonium sulphate by-product. For example, the Techni-Chem Process (1966) produces aminohexanoic acid from cyclohexanone with no major by-product (summarised in Chart 6).

The Toyo Rayon Co. have made hexanolactam by a photochemical process since 1963. Cyclohexane is converted directly to the cyclohexanone oxime by photochemical reaction with nitrosyl chloride (Chart 6). The oxime undergoes Beckmann rearrangement as in conventional processes. Ammonium sulphate production is cut to 2.2 tonnes per tonne of lactam. One important feature of this process is that it avoids the need for catalytic air oxidation of cyclohexane, which can give trouble, and the efficiency of conversion of cyclohexane is claimed to be very high ($>80\%$). At present this method seems to offer considerable potential for the production of cheap, high quality hexanolactam. But, there are no doubt problems involved in running a novel photochemical reactor.

Industrially, polymerisation of hexanolactam for fibre production is carried out either batchwise in an autoclave, or continuously in a column reactor. Both methods are slower than the corresponding processes for 6.6 nylon.

The autoclave process is very similar in that hexanolactam and water are heated under pressure, with a final period of gradual reduction of pressure of steam to atmospheric followed by extrusion under nitrogen. Because 6 nylon melts at about 220°C, polymerisation may be carried out at lower temperatures, and 260°C is usually the maximum.

Most 6 nylon is made by continuous polymerisation, an aqueous solution of lactam (with a little 6.6 salt or aminohexanoic acid as initiator) is pumped to the top of a vertical tubular reactor (the VK Tube) which may be 10 m high, a temperature gradient down the tube gradually heats the melt to 260°C as polymerisation occurs. Excess water boils off at the top, and molten polymer is removed at the bottom. The holding time in the reactor may be about 20 h— 6 nylon has much greater thermal stability than 6.6 nylon.

As for 6.6 nylon, acetic acid is used to limit relative molecular mass, and aqueous slurries of titanium(IV) oxide delustrant or of pigments may be added. Indeed, the addition of titanium(IV) oxide is easier than in 6.6 nylon technology for the problem of the aggregation of the dispersion of fine titanium(IV) oxide particles by the ionic 6.6 salt does not arise in lactam solutions.

The 6 nylon polymer contains about 10% by mass of free lactam under these conditions of polymerisation. This is the equilibrium quantity. It is removed before spinning otherwise it volatilises from the filaments as they emerge from the spinnerets, and makes the yarn sticky. A level of 0.5–1.0% is desirable, and is obtained by washing chip or by reducing the pressure above molten polymer to distil out the lactam. The rate of production of further lactam in the melt is sufficiently slow to allow removal and subsequent extrusion, and re-melting for spinning.

The equilibrium concentrations of the amine and carboxyl end-groups in molten 6 nylon (under steam) are very similar to those of 6.6 nylon, as one might expect. The numbers of amine ends and of carboxyl ends are equal in 6 nylon (in the absence of stabiliser). The number average relative molecular mass at equilibrium at atmospheric pressure in the absence of stabiliser is about 20 000 (and for 6.6 nylon also). This represents the maximum relative molecular mass of these polyamides which can be made normally at atmospheric pressures—to obtain higher relative molecular mass the pressure of steam above the molten polymer must be reduced.

6 nylon filaments are not exposed to steam at spinning, in contrast to the spinning of 6.6 nylon. If the freshly spun fibre is kept in dry air crystallisation causes shrinkage; absorption of water from the air causes an elongation. Stable cylinders of yarn are wound up under conditions of controlled humidity, and are held for a few hours before drawing to allow some crystallisation to occur.

FIBRE PROPERTIES AND USES

The greatest difference in properties between 6.6 and 6 nylons is the 50° difference in melting points. This imposes restrictions on the uses of the lower melting 6 nylon. There are other slight differences in the physical properties of the two fibres which give rise to some preferential choices for different end-uses. However, to avoid repetition, in a general treatment the two can be considered together and referred to without distinction under the generic name as 'nylon'.

The important point is the great versatility of nylon. Introduced (6.6 nylon) as a fine hosiery yarn for ladies' stockings, it rapidly took on a war-time role for parachute fabric and tow ropes. Since then its use has been extended into almost every conceivable area, and the development of new and improved yarns continues apace.

Versatility depends upon excellent basic properties, and the skilled recognition and exploitation of these properties. High strength, elasticity, and low density permit the manufacture of sheer stocking yarns: so successfully that the word 'nylons' is synonymous with ladies' hosiery. Its abrasion resistance is better than any other textile fibre, and large quantities of nylon are used in carpets, either as 100 % nylon or as blends with other fibres to improve the durability of the latter. Industrial uses depend also upon its toughness, fatigue resistance, and ability to absorb high velocity impacts (this because of its low glass transition temperature, coupled with its crystalline morphology). These uses include ropes, nets, slings, conveyor belts, and seat belts. The important use of 6.6 nylon in tyre cords has been mentioned in the previous chapter.

The drip-dry, no-need-to-iron properties of nylon have revolutionised apparel fabrics, and it is widely used in dresswear, lingerie, blouses, shirts, night attire, and socks. The textured yarns with increased cover, warmth, and softness have extended nylon's position in this area, and into related areas of bed-linen and furnishing fabrics. Bulked nylon staple fibres and blends are also used in hand knitting yarns. Yarns of non-circular cross-section provide lustrous fabrics, because of the different properties of light reflection, and better fibre cohesion and covering power. Carpet yarns may be roughly trefoil in section to give sparkle, and good packing.

The thermoplastic, crystalline character of nylon polymer means that fabrics can be heat set to size to give dimensional stability to subsequent washings or heat treatments, and to prevent permanent wrinkling or creasing during laundering.

Nylon yarns of different dyeing characteristics are available, as described earlier in the chapter, and in general nylon fabrics are easily dyed with a wide range of dyes.

The range of physical properties of nylon yarns is included in *Table 2.1* (Chapter 2) and may be from textile yarns of tenacity about 5 g dtex^{-1} at 30% extensibility and modulus 30 g dtex^{-1}, to industrial yarns of tenacity about 9 g dtex^{-1} at 16% extensibility and 40 g dtex^{-1} initial modulus. The higher tenacity is obtained by drawing to a higher degree, and is obtained at the expense of the extensibility—Chapter 3. The water vapour uptake of nylon is rather low at about 4% under standard conditions of measurement (65% relative humidity, 20°C), although greater than that of most other synthetic fibres.

A disadvantage of nylon, common to other synthetic fibres, is that it is easy to generate a static electrical charge which is not readily dissipated, especially at low humidities. This phenomenon is well known, giving rise to the cling of clothing to the body and frequently to electric spark discharges from clothing, or from finger tips after walking on a synthetic fibre carpet! Whilst much effort has gone into reducing the build-up of static charges on synthetic fibres, no really satisfactory general permanent solution to the problem is available yet. Static charges can be leaked from carpets by incorporating a small proportion of conductive fibres, for example metallised fibres or stainless steel fibres, or the new carbon coated 'Epitropic' fibres of I.C.I. Fibres Ltd.

The durability of nylon favours its use for synthetic recreational surfaces. Monsanto's 'Astroturf' was first used at the Houston Astrodome in 1966. Ribbon-like nylon fibres are used, backed by polyester or polyalkene fabric and foam rubber, and textured to suit athletics, football, etc.

OTHER POLYAMIDE FIBRES

Of the very large number of other possible polyamides and copoly-amides many thousands have been synthesised and tested in research laboratories throughout the world. Several have been selected for more detailed study and evaluation as fibres, but only a very few have so far reached commercial production as fibres. The reasons for the rejection of such large numbers have been outlined in Chapter 2.

Other polyamides are used for plastics applications as well as for fibres and combined outlets can add up to a worthwhile market, although small in comparison with the 6 and 6.6 giants.

The French Rhone-Poulenc Group spin nylon 11 under the name 'Rilsan'. Aminoundecanoic acid, $NH_2(CH_2)_{10}COOH$, monomer is obtained using castor oil as raw material to provide the 11-carbon atom chain: this dependence upon a natural product introduces an immediate check to production on an enormous scale at a reasonable and controllable price.

About 5×10^6 kg of fibres are produced, mainly for the European market, and socks and underwear are the principal end-uses because the fibre has a soft handle and does not irritate the skin. The moisture regain of nylon 11 is low, because of the greater proportion of hydro-carbon in the molecule, and this is of importance in some plastics applications.

It seems probable that nylon 12 will replace nylon 11 in the future because the 12-carbon atom unit can be synthesised economically now by combining together three molecules of butadiene. Nylons 12 and 11 are closely similar in physical properties, as is to be expected. The availability on an industrial scale of 12-carbon compounds derived from petroleum products enlarges the scope of polymer and fibre chemists. Du Pont have introduced a new polyamide fibre based upon the 12-carbon diacid, dodecanedioic acid, $HOOC(CH_2)_{10}COOH$. The diamine component is probably 4,4'-diaminocyclohexyl methane:

$$NH_2-CH \begin{matrix} CH_2-CH_2 \\ \diagdown \\ CH_2-CH_2 \end{matrix} CH-CH_2-CH \begin{matrix} CH_2-CH_2 \\ \diagdown \\ CH_2-CH_2 \end{matrix} CH-NH_2$$

This diamine is made from aminobenzene and methanal, followed by the hydrogenation of the benzene rings to cyclohexyl rings: it is potentially cheap, but known hydrogenation catalysts are very expensive. Three stereoisomers of the diamine are formed at the hydrogenation step. (These arise from the positions of the two substituents on each cyclohexyl ring relative to the general plane of the ring.) The relative proportions of the three isomers will affect the ways in which the polymer molecules can pack together, and hence polymer and fibre properties, and must be controlled. This is a sophisticated polymer structure. The presence of the rings of carbon atoms increases the stiffness of the molecule, and to counter this stiffness the more flexible 12-carbon acid has to be used. If the 6-carbon hexanedioic acid were used the polymer would be too high melting for a melt spinning process.

It is claimed that the fibres have greater resilience and are more like silk in handle than the established nylons, but it is too early to assess the future potential of this newcomer. Perhaps it is the first of the 'second generation' synthetic fibres which will eventually displace those of today, but so much depends upon the real worth of the improved properties and—most vital—upon the cost of the fibre. The textile industry and the customer will decide, but as yet it has the appearance of a speciality fibre for expensive fashion outlets.

Du Pont introduced another speciality fibre in 1961. Known as 'Nomex', the fibre has high temperature resistance and is non-flammable, with good mechanical properties; originally directed at military aircraft and space applications, a wider use in aircraft furnishing, protective clothing, electrical insulation, and filters has developed. Sewing threads which withstand the frictional heat of high speed machining are another use.

'Nomex' is prepared in solution and wet spun. The melting point is too high (475°C) for melt processing. The polymer is prepared by reacting 1,3-diaminobenzene with benzene-1,3-dicarboxyl chloride:

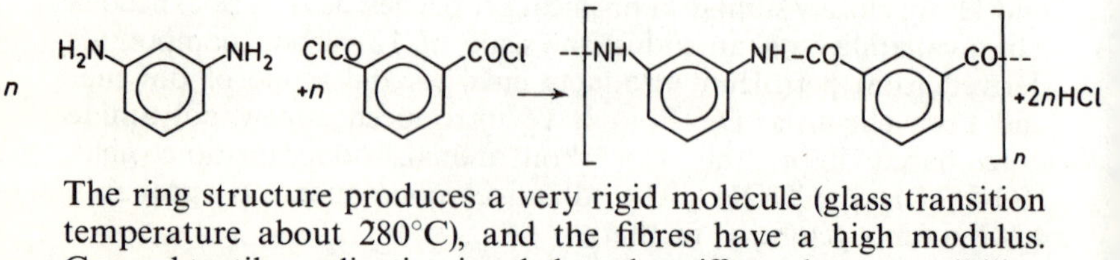

The ring structure produces a very rigid molecule (glass transition temperature about 280°C), and the fibres have a high modulus. General textile application is ruled out by stiffness, low extensibility, cost, and the limitation that the fibres are pale fawn or buff in colour.

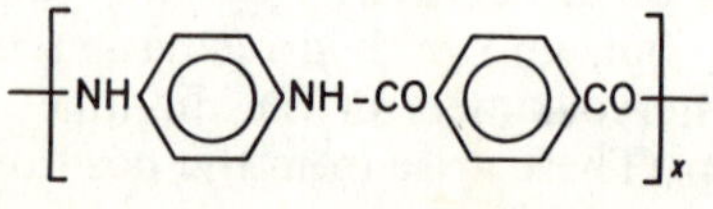

Du Pont have followed up this polymer with the development of the even more intriguing symmetrical structures based upon 1,4-diaminobenzene, and benzene-1,4-dicarboxylic acid (terephthalic acid, see following chapter), and upon the polymer from 4-aminobenzoic acid,

$$\left[NH\langle\bigcirc\rangle CO \right]_x$$

Both of these polymers are prepared by the acid chloride reaction, in the second case the chloride salt of the amino acid is converted to the acid chloride, and polymerisation is then carried out by dissolving in an amide solvent. No detail of the commercial preparation of these polymers is disclosed yet, but the chemistry is described in the patent literature. Fibres are being spun and evaluated under the general coding of 'Fibre B'.

The fibres are wet spun by dissolving the polymers in 100% sulphuric acid, or in sulphuric acid/SO_3, or in certain aliphatic amides. One would guess that sulphuric acid was a likely solvent for large scale use, with dilute acid or water in the spinning bath. The polymer molecules are very rigid, and cannot coil up on themselves like the flexible aliphatic polyamides, but rather exist in solution as extended rod-like entities. In the concentrated spinning dopes a liquid crystal or mesomorphic state exists. Here loose structural aggregations of parallel extended molecules are forming in the solution but without separation of a solid phase. In the simplest terms, this partial ordering in solution together with the orientation of the aggregates at spinning—that is, a flow orientation, rather like logs floating down a fast but non-turbulent river—gives a highly ordered extended chain structure in the solid fibre. The fibres are given a heat treatment to improve crystallinity and physical properties, but they are not drawn or stretched in the usual way, the oriented, anisotropic fibre structure is formed at spinning.

The fibres have high tenacity, and have been described with a modulus of the order of 10^{11} N m^{-2}, i.e. about 1000 g dtex^{-1}, which is approaching the modulus of carbon fibre (Chapter 11). The extensibility is of course low—a few per cent. On a weight basis, Fibre B has three or four times the strength of steel and a higher modulus. The commercial future of these materials as fibres will lie in tyre cords, and as reinforcing fibres for composite materials.

Polyamide plastics and fibres in general which contain aromatic ring systems offer improved properties, and many have been described in the scientific literature, but so far wider commercial development has been retarded because of the costly problems of synthesis and fabrication of high melting and difficultly soluble materials. Much work along these lines has been stimulated in the U.S.A. by the extremes of properties of new materials required for the space programme.

An elegant and novel use for synthetic fibres in the desalination of saline waters by the process of reverse osmosis was introduced by Dow Chemical Co. and by Du Pont in the mid-1960s. Their systems use hollow fibres assembled in bundles and sealed in a cylindrical pressure vessel, the open ends of the fibres pass through the seal. The walls of the hollow fibre act as a semi-permeable membrane, permitting the passage of water but not of dissolved salts. Accordingly, if a saline solution is pumped under pressure greater than its osmotic pressure to the cylindrical vessel, water which is much reduced in salt content will penetrate the fibres and flow along the hollow centres, whence it may be collected outside the seal. The fibres themselves, as fine capillaries, withstand the external pressure without collapse. Dow units are based upon cellulose acetate fibre technology, but Du Pont use hollow aromatic polyamide fibres, possibly copolymers of the Fibre B type. The use of hollow 6.6 nylon fibres is also described by both companies. In a variation of this membrane separation technology, Du Pont use hollow polyester fibres for gas separations by multiple diffusion methods. The development and use of these separation techniques in applications as diverse as seawater desalination and artificial kidney machines is expected to grow in the future.

4 NYLON

Yarns spun from this polymer are not yet commercial, only pilot plant quantities are being made for evaluation in the U.S.A., where production may begin later this decade. However, 4 nylon deserves mention for it offers some interesting chemistry and fibre properties, and its discovery and development provides an exceptional modern example of persistence and scientific thoroughness (the story is told in the journal *Chem. Tech.*, January, 1972 issue).

The monomer is pyrrolidone, the cyclic lactam of 4-aminobutanoic acid. The stable 5-membered ring cannot be polymerised

$$
\begin{array}{ccc}
CH_2 & - & CH_2 \\
| & & | \\
CH_2 & & NH \\
\backslash & & / \\
& CO &
\end{array}
$$

hydrolytically, in the same way that hexanolactam is polymerised, and was regarded as unpolymerisable. In 1951 C. E. Barnes and W. O. Ney (U.S.A.) found that very strong bases such as alkali metals, and their hydrides or hydroxides, would bring about polymerisation in the presence of various other compounds which act as co-initiators. The discovery arose accidentally following a careful investigation into the apparently erratic behaviour of the addition of ethyne to pyrrolidone catalysed by alkalis. Critical though it was, this discovery did not permit exploitation of the polymer for the relative molecular mass was low and it depolymerised rapidly to pyrrolidone at its melting point (about 260°C). (Compare the hexano-lactam—6 nylon equilibrium.)

A great deal of effort went into searches for means of stabilising the molten polymer, but without success. Barnes took up the problem again in 1967, working privately in his own laboratory, and by carefully following through observations he unexpectedly found that alkali initiated polymerisation in presence of carbon dioxide as co-initiator gave polymer of higher relative molecular mass, with a narrow distribution of molecular masses, and, most important, of greatly improved thermal stability. It is claimed that this 4 nylon is sufficiently stable to be melt spun, provided that water is excluded, and that excellent textile fibres are obtained.

Fibres of 4 nylon are reported to have a similar range of physical properties to those of 6 nylon, with the important exception that its moisture regain at 65% relative humidity is higher than that of cotton. This is very significant in that for the first time we have a synthetic with the toughness and durability of nylon coupled with high uptake of water vapour. Many problems must remain to be solved before the technical and commercial potential of 4 nylon can be assessed, but if it is successful it must make inroads into some of the markets held by cotton.

The mechanism of polymerisation is anionic; the usual laboratory initiator is potassium hydroxide, which forms the potassium salt of pyrrolidone and water (which must be distilled off). The pyrrolidone anion then reacts with the carbamate formed by reaction of CO_2 and pyrrolidone:

Polymerisation proceeds by repeated addition of the anion to the carbonyl of the cyclic N-acyl amide end-unit.

It should be noted that hexanolactam can be readily polymerised by a similar mechanism, using N-acyl amides or related compounds as the co-initiators, but the anionically polymerised material is not used for fibres. Several side reactions can occur, and control of the polymer end-groups is much easier with the hydrolytically polymerised material.

7

Polyester Fibres

'Terylene' is the registered trade mark of I.C.I.'s polyester fibre, a fibre which was a British invention and which is currently the most rapidly growing in use of the synthetic fibres. In 1971 world production of all polyester fibres was about 2×10^9 kg, in comparison with polyamide fibres at the same level: five years earlier the figures were 0.6×10^9 and 1.2×10^9 kg respectively (*Figure 1.1*).

W. H. Carothers made polyesters during the course of his pioneering work by reacting aliphatic diols with aliphatic dicarboxylic acids. He obtained low melting solids from which fibres could be melt spun, but were too low melting to offer any encouragement for making useful fibres. Carothers turned his attention, with success, to the higher melting polyamides. J. R. Whinfield and J. T. Dickson, in the laboratories of the Calico Printers Association in Manchester, investigated the effects of introducing aromatic dicarboxylic acids and thereby 'stiffening' the polyester molecules (Chapter 2).

The polyester from 1,2-ethane diol (ethylene glycol) and benzene-1,4-dicarboxylic acid (terephthalic acid) was prepared and found to melt at about 265°C. Further, excellent fibres could be made from the melt. The polyester is commonly known as poly(ethylene terephthalate), and the reaction can be represented as:

$$n\,HO(CH_2)_2OH + n\,HOOC\text{—}\langle\bigcirc\rangle\text{—}COOH \longrightarrow$$

$$H\text{—}\!\left[O(CH_2)_2OCO\text{—}\langle\bigcirc\rangle\text{—}CO\right]_n\!\!\text{—}OH + (2n-1)H_2O$$

A patent was taken out in 1941. War delayed exploitation, but then I.C.I. in Britain, and Du Pont in the U.S.A. acquired the patent rights and developed industrial processes which led to production in 1953.

In contrast to the introduction of nylon, trade names were registered for the individual fibres. (The Du Pont fibre is called 'Dacron'.) In this way, one advertises specifically one's own fibre, and the customer is sure of what quality he is buying.

Today poly(ethylene terephthalate) fibres are produced in many countries throughout the world, under a variety of trade marks, and practically all fabrics and garments classed as 'polyester fibre' are of this polymer.

In the 1940s ethanediol was already an industrial chemical, with large scale use in antifreeze coolants, but benzene-1,4-dicarboxylic acid was only a laboratory chemical. Large scale processes for manufacture of the acid had to be devised and put into operation.

Ethanediol is obtained from ethene which is in abundant supply from the thermal cracking of oil. The older industrial process reacts ethene with chlorine and water under pressure to form 1-chloro-2-hydroxyethane (chlorohydrin),

$$CH_2{=}CH_2 + Cl_2 + H_2O \longrightarrow ClCH_2CH_2OH + HCl$$

which is then hydrolysed with lime slurry or with sodium hydroxide solution to 1,2-epoxyethane (ethylene oxide),

$$ClCH_2CH_2OH + OH^- \longrightarrow CH_2{-}CH_2 + Cl^- + H_2O$$
$$\diagdown O \diagup$$

A more modern process oxidises ethene directly to 1,2-epoxyethane using air and a silver catalyst at 250–325°C and 1–3 MPa. Direct oxidation avoids the use of chlorine (which is converted to a by-product without value), but is complex with an expensive catalyst, and is reported to need careful control—some of the ethene is oxidised through to carbon dioxide.

The epoxyethane is converted to the diol by reaction with water at about 180°C,

$$CH_2{-}CH_2 + H_2O \longrightarrow HOCH_2CH_2OH$$
$$\diagdown O \diagup$$

The principal route to terephthalic acid is by oxidation of 1,4-dimethylbenzene (p-xylene). Distillation of mineral oil produces a fraction which consists of alkyl-substituted cycloalkane hydrocarbons. Treatment with hydrogen under pressure at about 500°C, in contact with molybdenum(VI) oxide and alumina catalysts, dehydrogenates the cycloalkanes to aromatics. Fractional distillation of the mixed aromatics gives a 'xylenes fraction' which, typically, is composed of:

1,3-dimethylbenzene	45%	(Bp 139°C, mp −48°C)
1,2-dimethylbenzene	22%	(Bp 144°C, mp −25°C)

1,4-dimethylbenzene 20% (Bp 138°C, mp 13°C)
ethylbenzene   13% (Bp 136°C, mp −95°C)

The boiling points are too close together for economic separation by fractional distillation, but the higher melting 1,4-dimethylbenzene can be separated by crystallisation from the cooled liquid. Several stages are needed to obtain the maximum yield of the isomer at better than 98% purity. The proportion of 1,4-dimethylbenzene in the 'xylenes' fraction is improved by isomerising the ortho and meta isomers over metal oxide catalysts, and the aromatisation and isomerisation stages can be combined.

There are several processes for oxidation of 1,4-dimethylbenzene to acid. The route first developed by I.C.I. and by Du Pont uses nitric acid as oxidising agent (about 30% acid at 180°C or so). The product is contaminated with some nitro-substituted acids.

$$CH_3 \langle \bigcirc \rangle CH_3 \xrightarrow{\text{Oxidn}} HOOC \langle \bigcirc \rangle COOH$$

Liquid phase catalytic air oxidation processes have been devised, and are now widely used. Cobalt(II) salts, or manganese(II) salts activated by bromine, are employed as catalysts in acetic acid solution. Oxidation of the first methyl group to carboxylic acid is easier than oxidation of the second, hence 4-methylbenzenecarboxylic acid (p-toluic acid) is the main impurity.

Purification of benzene-1,4-carboxylic acid (especially from nitro-acids) is difficult because it is almost insoluble in commonly available solvents. To reach the exacting standards of purity required for polymerisation, it is usual to convert it to the dimethyl ester. Direct catalysed esterification with methanol is slow because of the insolubility of the acid, and hence is carried out under pressure at high temperatures. Dimethylbenzene-1,4-carboxylate is then purified by low pressure distillation. The pure ester melts at 140.8°C.

$$HOOC \langle \bigcirc \rangle COOH + 2CH_3OH \longrightarrow CH_3OCO \langle \bigcirc \rangle COOCH_3 + 2H_2O$$

1,4-dimethylbenzene is also oxidised and converted to dimethyl-benzene-1,4-carboxylate in a four stage process (the Witten process), which has been adapted to continuous operation, and which gives good yields. Cobalt catalysed air oxidation of 1,4-dimethylbenzene produces 4-methylbenzenecarboxylic acid,

$$CH_3 \langle \bigcirc \rangle CH_3 \xrightarrow[\text{Co}]{O_2} CH_3 \langle \bigcirc \rangle COOH$$

which is converted to its methyl ester,

$$CH_3\langle\bigcirc\rangle COOH + CH_3OH \longrightarrow CH_3\langle\bigcirc\rangle COOCH_3 + H_2O$$

The second methyl group is then oxidised using a copper catalyst, and the product fully esterified in the final step:

$$CH_3\langle\bigcirc\rangle COOCH_3 \xrightarrow[Cu]{O_2} HOOC\langle\bigcirc\rangle COOCH_3 \xrightarrow[CH_3OH]{} CH_3OOC\langle\bigcirc\rangle COOCH_3$$

This process is used in Europe and in the U.S.A. The economics of the Witten process are said to be very similar to the economics of direct air oxidation, followed by esterification, and superior to the older nitric acid process.

There is no need to hydrolyse pure dimethylbenzene-1,4-dicarboxylate to obtain pure acid for subsequent reaction with diol. Polymerisation direct from the ester is carried out in two stages. In the first step, methanol is eliminated by ester exchange with a small

$$CH_3OOC\langle\bigcirc\rangle COOCH_3 + 2\ HOCH_2CH_2OH \longrightarrow$$
$$HOCH_2CH_2OOC\langle\bigcirc\rangle COOCH_2CH_2OH + 2\ CH_3OH$$

excess of ethane diol, Reaction is carried out at the boiling point of the ethanediol (198°C) and methanol distils off. A catalyst is used. Many are described in the patent and scientific literature, but few are completely satisfactory in practice. Various metallic oxides and salts, antimony(III) oxide, and germanium compounds can be used. The problems arise at a later stage, because the catalyst may discolour the final polymer and adversely affect its thermal stability.

All traces of methanol must be removed in the shortest possible reaction time, for residual methanol will act as a stabiliser and limit the ultimate attainable relative molecular mass. Some polymerisation occurs during this first stage, but only to low relative molecular mass.

Titanium(IV) oxide delustrant dispersed in diol is now added (if desired), and the first stage product heated in a stirred autoclave, where polymerisation is completed at 280°C and at a pressure of about 130 Pa (1 mm Hg). Ester exchange reactions continue, with liberation of ethanediol which distils off at the low pressure:

$$2 \ HOCH_2CH_2OOC \langle \bigcirc \rangle COOCH_2CH_2OH \longrightarrow HOCH_2CH_2OH +$$

$$HOCH_2CH_2OOC \langle \bigcirc \rangle COOCH_2CH_2OOC \langle \bigcirc \rangle COOCH_2CH_2OH$$

$$H \left[OCH_2CH_2OOC \langle \bigcirc \rangle CO \right]_n OCH_2CH_2OH +$$

$$H \left[OCH_2CH_2OOC \langle \bigcirc \rangle CO \right]_m OCH_2CH_2OH \longrightarrow$$

$$HOCH_2CH_2OH + H \left[OCH_2CH_2OOC \langle \bigcirc \rangle CO \right]_{(n+m)} OCH_2CH_2OH$$

The catalyst functions for both stages. Oxygen has to be excluded, of course, and the whole polymerisation is carried out under nitrogen.

Removal of ethanediol is facilitated by stirring, and the course of polymerisation can be followed by monitoring the torque on the stirrer shaft (a measure of the melt viscosity). When the desired extent of reaction is achieved, the molten polyester is extruded from a slot, or from holes, in the bottom of the reactor and quenched in water, or on the surface of a water-cooled drum. The ribbon or lace is then cut to chip, much as for polyamides.

The above polymerisation sequence has been adapted to a continuous process. High relative molecular mass polyester, which is needed for high tenacity yarns, is not easy to make in batch reactors because the time at high temperature is long enough to allow degradation reactions to have effect. Continuous polymerisation processes generally make better polymer with less time at high temperatures. (For example, it takes time to empty a large autoclave of viscous polymer.) The main degradation reactions which limit the degree of polymerisation are believed to be as follows:

$$--- OC \langle \bigcirc \rangle CO.OCH_2CH_2O.OC \langle \bigcirc \rangle CO ---$$

$$\downarrow$$

$$--- OC \langle \bigcirc \rangle CO.OCH = CH_2 + HOOC \langle \bigcirc \rangle CO ---$$

$$\downarrow$$

$$---OC\langle\bigcirc\rangle CO.O.CO\langle\bigcirc\rangle CO--- \; + CH_3CHO$$

$$---OC\langle\bigcirc\rangle CO.OCH=CH_2 \rightarrow ---OC\langle\bigcirc\rangle COOH + CH\equiv CH$$

and other products. Some of these reactions may be promoted by the ester exchange catalysts.

Whilst dimethylbenzene-1,4-dicarboxylate offers advantages in purification, every kilogramme of ester provides only 0.86 kg of acid for use in polymerisation, the remainder is recovered as methanol. Hence there is some economic attraction in making purified 'fibre-grade' benzene-1,4-dicarboxylic acid for direct use. Acid made by catalytic air oxidation of 1,4-dimethylbenzene can be purified to a sufficient quality, and processes have been introduced on an industrial scale. The purified free acid is a little more costly to make than the pure ester, so the overall polymer cost when using free acid is said to be less than 10% cheaper. This is not enough to cause manufacturers to abandon established practice and equipment, but future new plant may increasingly use free benzene-1,4-dicarboxylic acid technology.

Polyester chip must be dry before melt spinning, otherwise hydrolysis of the ester groups will occur and the relative molecular mass will fall. Spinning is carried out under dry nitrogen at about 290°C. The filaments are amorphous when wound up. The polymer cannot crystallise below the glass transition temperature (about 90°C), and this is not sensitive to water like the polyamide 6.6 nylon. The amorphous filaments are stable on the wind up cylinder.

The continuous filament has to be hot drawn. The drawing process introduces molecular orientation and allows crystallisation of the polymer. The hot filaments cool before drawing tension is released, and some of the strain is frozen in and will be recovered by shrinkage when the yarn is next heated. Thus, a heat relaxation treatment is given to provide dimensional stability in the finished textile fabric.

Polyester fibres have excellent physical properties and accordingly very wide-ranging uses. The high modulus of elasticity (over 100 g dtex^{-1}, and surpassed only by flax), high glass transition temperature, and insensitivity to water, all contribute to the resilient, warm handle of fabrics with high dimensional stability and resistance to accidental creasing, but with the ability to retain heat-set creases (Chapter 2). The abrasion resistance is high, although not as great as nylon.

These physical properties make polyester fibres excellent for use in fashion outerwear. Textured continuous filament fibres, familiar under I.C.I.'s 'Crimplene' registered trade name, have popular use in knitted and woven dress and suit fabrics. Staple fibre is blended with wool for worsted suitings, with cotton for shirts, dresses, underwear, and light weight suitings, and with flax for crease-resistant linen.

The resistance of polyester to light, i.e. to photo-initiated oxidation, is very good, and the fibre can be used for curtains, and for outdoor sportswear, tents, marquees, and sails.

You will not find ladies' tights and stockings of polyester, for the elastic modulus is too high. Tights have to freely accommodate extensive movements, and must recover rapidly without 'bagging' at the knee. Nylon has the better elastic properties and abrasion resistance for this purpose. However, the high modulus and resilience of polyester allows its use as a filling fibre for quilts, pillows, and sleeping bags (together with the great advantage that the complete article can be laundered), and as a carpet fibre.

High tenacity 'poly(ethylene terephthalate)' fibres find uses as diverse as sewing threads, ropes, and tyre cords. The excellent light stability is an added attraction for the use of polyester in nets and ropes. Polyester as a tyre cord provides a higher modulus than nylon, and better dimensional stability. Because of the higher glass transition temperature it is less prone to creep when the tyre is hot from running. Good dimensional and light stability, together with impact resistance, are exploited in car seat belts, conveyor belts, and transmission belts.

General chemical inertness, especially resistance towards hydrolysis by acids, coupled with toughness, favour application in fabrics for filters, for laundry bags, and for paper makers' felts (which are used to convey wet, acidic paper to dryers). Fabrics coated with rubber, PVC, and other plastics, are used for hoses, protective covers, and containers.

As a complete contrast, this chemical inertness permits medical uses within the body for sutures, and for blood vessel prosthetics. Knitted, transversely corrugated tubes are used to replace sections of blood vessels. The permeable fabric structure allows penetration and clotting, and then interpenetration and overgrowth of tissue.

OTHER POLYESTER FIBRES

No other polyester fibre compares with 'poly(ethylene terephthalate)' in volume of production. One other benzene-1,4-dicarboxylate polyester fibre is made in the U.S.A. and is called 'Kodel'. This is the trade name of Eastman Kodak's fibre which is believed to be the melt spun polyester made from benzene-1,4-dicarboxylic acid and 1,4-bis(hydroxymethyl) cyclohexane,

$$HOCH_2CH \begin{matrix} CH_2-CH_2 \\ \\ CH_2-CH_2 \end{matrix} CHCH_2OH$$

This diol has two isomeric *cis* and *trans* forms. The *trans* isomer gives more symmetrical polymers with the higher melting points. A mixture of about 70% *trans* and 30% *cis* diols gives a polybenzene-1,4-dicarboxylate ester melting at about 290°C, so this is a possible composition for the fibre.

The Japanese are producing a new fibre which is said to be like silk in physical properties and in handle. Made by the Nippon Rayon Company, the newcomer is called 'A-Tell'. Little chemical detail is published yet, but it is an aromatic ester-ether polymer probably prepared by reacting 4-hydroxybenzenecarboxylic acid with ethanediol, or with 1,2-epoxyethane, and has the structure:

$$H \left[OCH_2CH_2O \left\langle \bigcirc \right\rangle CO \right]_n OH$$

It is thus very similar to 'poly(ethylene terephthalate)', but has a melting point of about 225°C. The fibre is being used for Japanese and European style woven fabrics, and for knitted dress fabrics. There will doubtless be wider application as the fibre becomes more familiar and more readily available. The rather low melting point may restrict development in some industrial fabrics.

The polyester obtained by the self-condensation of 4-hydroxy-benzenecarboxylic acid has been prepared by The Carborundum Company of the U.S.A. and is marketed under the registered trade name 'Ekonol'.

$$H \left[O \left\langle \bigcirc \right\rangle CO \right]_n OH$$

The free hydroxy-acid will not polymerise on heating. It is unstable, and decarboxylates; the method of preparing 'Ekonol' is not disclosed yet, but possibly a route via the acid chloride is used (as for the analogous polybenzamide, Chapter 6). In addition to the difficulties of making it, the polyester presents formidable fabrication problems. It decomposes at about 550°C without melting, and no solvent is known. It can be formed by a compression sintering process at 425°C under pressure. We are cheating in describing this polyester, because so far no one has disclosed fibres made from it.

It remains a fascinating polyester challenger to our ingenuity!

One new fibre remains which is not a polyester, but fits in here perhaps better than elsewhere. This material also is made by The Carborundum Company and is called 'Kynol'. The polymer is an undisclosed cross-linked phenolic structure—possibly analogous to the phenol–methanal plastics. It is, therefore, also insoluble and infusible. Probably fibres are spun in some way from a linear polymer, and then the reaction and cross-linking completed in the fibrous form. When exposed to a flame, the fibre does not melt, and does not burn, but carbonises with very little smoke. It is being used to make protective clothing, fire-proof furnishings, filter cloths, and as a heat insulating fibre—for the density is lower than glass.

8

Acrylic Fibres

Cyanoethene (acrylonitrile, $CH_2\!\!=\!\!CH.CN$), is a relatively cheap chemical which polymerises to a high softening point polymer from which textile fibres with a soft, wool-like handle may be spun. The early development of commercial fibres was retarded by the intractable nature of the polymer. It is insoluble in common organic solvents and on heating to its softening point does not melt, but undergoes chemical changes and darkens in colour. The fibres are not easy to dye, and their low thermoplasticity makes some fibre processing operations difficult.

These problems were overcome (1) by a search for suitable solvents to give polymer solutions from which fibres could be spun, and (2) by copolymerising up to 10% by mass of additional polar monomers with the cyanoethene. The comonomers break up the regularity of the polymer structure and so increase its solubility and thermoplasticity, and make it more easily penetrated by water and dyes.

The most usual comonomer is methyl propenoate (methyl acrylate, $CH_2\!\!=\!\!CHCOOCH_3$) but basic or acidic monomers may be included also to give affinity for specific classes of dyestuffs. The actual comonomers used remain each fibre producer's secret, but the facility of copolymerisation offers scope for a wide variety of fibre properties.

The first commercial fibres were introduced in 1948 by Du Pont ('Orlon') and by Baeyer ('Dralon'). Dimethylmethanamide (Dimethylformamide) $[HCON(CH_3)_2]$ was used as solvent, and fibres may be spun using the dry spinning technique, or wet spun by extruding into an aqueous dimethylmethanamide coagulating bath. Wet spinning is favoured for staple fibre production.

Several fibres based on poly(cyanoethene), better known under the trivial name of poly(acrylonitrile), are produced now throughout the world: in 1969 production totalled about 9×10^8 kg, 1971 production capacity was estimated as about 15×10^8 kg. The generic term 'Acrylic' is applied to all fibres which are composed of more than 85% cyanoethene.

The industrial routes to cyanoethene are summarised in Chart 7. Early processes were based on addition of hydrogen cyanide to ethyne, or alternatively to epoxyethane, or to ethanal. The most attractive modern routes, introduced about 1960, begin from propene—a cheap tonnage product of the petroleum industry. The Sohio process, as an example, employs a catalytic air oxidation of propene in the presence of ammonia for the direct production of cyanoethene (Chart 7). Bismuth phosphomolybdate is used as catalyst. The capital costs are low, and in the U.S.A. from propene at 4–5 ¢ kg^{-1} the raw materials' cost to produce cyanoethene is said to be only 8 ¢ kg^{-1}. This low price potential of the monomer will ensure an increasing production and use of the fibre.

(1) $CH\equiv CH + HCN \xrightarrow[Cu^+/NH_3]{} CH_2=CHCN$

(2) $CH_2\!-\!CH_2 + HCN \longrightarrow HOCH_2CH_2CN$
O
$CH_2=CHCN + H_2O$

(3) $CH_3CHO + HCN \longrightarrow CH_3\!-\!CHCN \longrightarrow CH_2=CHCN + H_2O$
OH

(4) *SOHIO PROCESS*

$2NH_3 + 3O_2 + 2CH_2=CHCH_3 \xrightarrow[450°C]{Bi\ phospho-molybdate} 2CH_2=CHCN + 6H_2O$

200–300 kPa

Chart 7 *Industrial routes to cyanoethene*

Cyanoethene monomer is polymerised by a free-radical initiator (Chapter 2), by repetitive addition of the double bonds of monomer molecules:

$CH_2=CH.CN + R^{\cdot} \longrightarrow R.CH_2-\dot{C}H$
CN

$R.CH_2-\dot{C}H + n\ CH_2=CH.CN \longrightarrow -(CH_2\!-\!CH)_x-$
$CN \qquad\qquad\qquad\qquad CN$

Polymerisation may be carried out in solution in dimethylmethan-amide, or in suspension in water (the monomer is reasonably soluble in water), and continuous polymerisation processes have been developed. Polymerisation in water allows better control. Hydrogen peroxide, peroxydisulphates, or redox catalysts are used as radical sources at 30–70°C. The suspended particles of polymer are filtered, washed, and dissolved in the spinning solvent.

Whilst poly(cyanoethene) yarns are spun as continuous filament, very little is used as such, by far the greater quantity is wet spun and converted to staple fibre. The spun yarns are stretched in hot air at about 100°C, or more usually in steam or in hot water (70–100°C). The dry polymer has a glass transition temperature about 80°C which is depressed by water, and stretching must be carried out above this temperature. High tenacity fibres made by a high degree of stretch tend to fibrillate. The tensile properties are similar to cotton (*Table 2.1*), but the extensibility is greater, and the moisture regain is low (about 1.5% water at 65% R.H. and 20°C).

The physical properties are little affected by cold water, as would be predicted, but in hot water the glass transition is exceeded and the acrylic fibres show plasticity, and will flow under load. Perhaps this is of no consequence if fashion dictates a long slim line, but otherwise is a disadvantage in laundering!

The polymer has a highly ordered structure in an oriented fibre, but only in a lateral direction, there is no evidence for a regular molecular arrangement along their lengths. It cannot be described as crystalline. This unusual structure probably gives rise to the lack of length stability of acrylic fibres in hot, wet conditions (i.e. above the glass transition temperature), but restricts penetration by dyes.

However, the defect of lack of stability can be turned to good use in making excellent bulked fibres. If fibre is hot stretched and cooled whilst held to length before cutting to staple, on re-heating it will shrink, the molecules trying to return to their pre-stretch, stress-free state. By mixing high shrinkage staple with normal staple, spinning to yarn, and heating wet to produce shrinkage, a soft, fluffy, high bulk yarn is obtained which is excellent for knitting. The staple component which does not shrink is forced into loops and waves to accommodate the shrinkage. Bicomponent acrylic fibres are spun with two polymers of differing shrinkage (slightly different copolymer compositions) and give bulked staple on shrinking (Chapter 12).

As a cheap, predominantly staple fibre product, acrylic fibres are used in blends with other staple fibres, natural and man-made: blends with wool are especially attractive, and are used for suits, and skirts. Acrylic blends with cellulosic fibres are used for dresses, underwear, and so on. However, much 100% acrylic yarn (especially bulked) is also used for fabrics and for knitting. Knitwear is an important outlet for acrylics. The abrasion resistance is reasonable and resilience is good, so the fibres are used in carpets. The soft, warm handle has led to their use in blankets, and in pile fabrics.

Dyeing was difficult initially, but development of copolymers has eased the problem and a good range of colours and depth of shade can now be applied to acrylics and to the various blends.

MODACRYLIC FIBRES

If more than 15% by mass of comonomer is copolymerised with cyanoethene the properties of the polymer are more drastically altered and quite different fibres obtained. A new generic name, 'Modacrylic fibre' (i.e. modified acrylic), is given to such fibres which contain less than 85% but at least 35% by mass of cyanoethene. Such a severe break up of the regularity of the poly(cyanoethene) structure produces materials which dissolve in a greater variety of solvents, and which are thermoplastic with more definite melting points. Spinning becomes less of a problem in some ways.

Of the large number of comonomers which could be selected, in practice only two are used on a commercial scale, chloroethene (vinyl chloride, $CH_2{=}CHCl$), and 1,1-dichloroethene (vinylidene chloride, $CH_2{=}CCl_2$).

The first modacrylic was introduced by Union Carbide (U.S.A.) in 1948, and called initially 'Vinyon N'. Of composition 40% cyanoethene and 60% chloroethene, the fibre is now sold as staple under the name 'Dynel'.

The modacrylic fibres are spun from solution in propanone by either wet or dry processes. As with the acrylics, the great bulk is used as staple fibre, and the economics of staple production favour wet spinning. The fibres are stretched and heat annealed. The heat treatment improves the 'crystallinity' (such as it is), and relaxes out strains, because, as one would guess, the dimensional stability is not good, especially under hot, wet conditions. Yarns with high shrinkage can be produced.

Tensile properties depend upon the degree of stretch and heat treatment, but commercial fibres are much less stiff than acrylics (as is to be anticipated) with a modulus of elasticity less than half that of an acrylic fibre. Modacrylics have relatively small speciality markets: production is in terms of 10^7 kg per annum—about one hundredth that of the acrylics. The fibres have good resilience, and abrasion resistance comparable to cotton, but their thermoplasticity is too great for general apparel textile markets. Their main distinguishing feature is that they will not propagate a flame, and shrink away from it: if the flame is continuously applied the fibre will burn. Hence modacrylic fibres find use in curtains and upholstery in public buildings and ships where there is a serious fire hazard. But, the greatest quantity is used in deep-pile fabrics, either alone or blended (especially with acrylics). These deep-pile fabrics are used as linings for coats, boots, etc., and as outerwear for coats, collars and trimmings, and hats. Furs, rugs, and carpets provide other outlets, and there are some industrial applications. Two rather curious, but significant, uses are for paint roller covers and in wigs and hair pieces—established items of contemporary life.

CHLOROETHENE FIBRES

The flame retardant property of modacrylic fibres is a highly desirable characteristic in many textile applications, it arises from the presence of chlorine in the comonomer together with high shrinkage from the flame. Other polymers and copolymers of chloroethene and 1,1-dichloroethene may be expected to have this property to some degree and are indeed produced, and also find markets as speciality fibres. In fact the very first synthetic fibre on a commercial scale was a chlorinated poly(chloroethene) produced in Germany in 1934, and still manufactured there (PeCe, by I. G. Farbenindustrie).

The reason for treating poly(chloroethene) with more chlorine is to make it soluble in propanone for, in common with poly(cyanoethene), the polymer initially presented problems of insolubility in cheap solvents, and it cannot be melt spun. Chlorination is carried out to increase the chlorine content from about 57% to 64%.

The French Company Rhodiaceta discovered that poly(chloroethene) fibre could be dry spun from mixtures of propanone and carbon disulphide (1941), and have continued to produce this fibre since that time. The attraction of chloroethene is its cheapness, as it is produced in very large quantity for the plastics industry—more familiar, perhaps, as poly(vinyl chloride) abbreviated to PVC.

The monomer is made by reacting ethene with chlorine to form 1,2-dichloroethane, and then eliminating HCl by heating in the presence of catalysts.

$$CH_2{=}CH_2 + Cl_2 \longrightarrow CH_2Cl.CH_2Cl$$
$$CH_2Cl.CH_2Cl \longrightarrow CH_2{=}CHCl + HCl$$

The chief limitation in properties of poly(chloroethene) fibres is the low softening point (about 75°C) above which the fibres shrink: they are also sensitive to chlorinated dry-cleaning solvents. This shrinkage of course is turned to advantage to make novelty bulk and patterning effects in fabrics, and in moulded garments, but limits any wide application in textiles.

Low flammability, together with good resistance to attack by aqueous acids and alkalis (the fibres do not take up water at all), and resistance to sunlight, has ensured speciality uses in upholstery, theatre stage curtains, protective clothing, flying suits, insulation, and filter cloths.

Recently, improved forms of poly(chloroethene) fibres have been developed by radical polymerisation at low temperatures (perhaps 0 to $-70°C$). The polymer so produced has a higher softening point and improved resistance to dry cleaning solvents, these improved properties almost certainly arise from an increased stereo-regularity of the polymer with resultant improvement of molecular packing in the fibre. So far, reported changes in fibre properties are significant but not dramatic.

Copolymers of 1,1-dichloroethene with chloroethene are melt spun to fibres. Again, poly(1,1-dichloroethene) is soluble only with difficulty, and melt spinning of the homopolymer is possible but very difficult. The copolymers themselves have only a limited stability at the melt temperature (about $180°C$), hydrogen chloride is eliminated and the decomposition is catalysed by several metals— including iron. Extruders of nickel or cobalt alloys are usually employed, with short melt holding times.

Copolymers of 1,1-dichloroethene used for fibres are crystalline, but soften at about $140°C$, which is too low for general textile end-uses. The markets for the fibres are similar to those for the PVC fibres.

POLY(ETHENOL) FIBRES

A completely different type of ethenyl (vinyl) fibre has been extensively developed by the Japanese, this is poly(ethenol)—or poly(vinyl alcohol)—and is known as 'Vinylon' in Japan. The intriguing property of poly(ethenol) is that it is soluble in water: it is a tribute to Japanese chemistry and technology to produce a fibre of use in fishing nets from a water-soluble polymer! A second peculiarity is that ethenol monomer is too unstable to be isolated (ethanal is the stable isomer), and poly(ethenol) is prepared by hydrolysis of polymerised ethenyl acetate monomer.

Ethenyl acetate is prepared by mercury salt catalysed addition of acetic acid to ethyne:

$$CH{\equiv}CH + CH_3COOH \xrightarrow{\text{Hg}} CH_2{=}CH.O.CO.CH_3$$

The purified monomer is polymerised by free radical initiation and the polymeric ester is then hydrolysed to poly(ethenol) with alcoholic sodium hydroxide.

$$
\begin{array}{c}
\text{---CH}_2\text{---CH---CH}_2\text{---CH---} \\
| \qquad\qquad | \\
\text{OCO.CH}_3 \quad \text{OCO.CH}_3 \\
\downarrow \\
\text{---CH}_2\text{---CH---CH}_2\text{---CH---} \\
| \qquad\qquad | \\
\text{OH} \qquad\quad \text{OH}
\end{array}
$$

Poly(ethenol) fibres are wet spun from water into a coagulating bath of sodium sulphate solution. The fibre is made insoluble in water by a hot stretch which gives high molecular orientation and crystallinity; resistance to hot water is further improved by reaction with methanal. The methanal reacts with some of the hydroxyl groups—probably those which are not in the crystalline regions and therefore are more easily accessible and converts them to ether links with less affinity for water. Some ether cross-links are probably formed, but the most important factor in making the fibre insoluble is the crystallinity, and not covalent cross-links. In one sense the fibre is a 'synthetic cellulose', packed full of hydroxyl groups, and the moisture regain is about 5%, which is very high in comparison with other ethenyl polymers, but low considering the water-solubility of the parent polymer.

The physical properties are superior to those of rayon, especially in elasticity and abrasion resistance, which are comparable with the polyamide and polyester fibres, and high tenacity yarns (8.5 g dtex^{-1}) of high initial modulus (180 g dtex^{-1} maximum quoted value, about the same as flax) can be obtained, but these values are decreased when the fibre is wet. The wet strength is about 80% of the dry value. Dyeing is not as easy as with cellulosic fibres.

The fibre softens at about 200°C, which is rather low, but is said to have fairly wide application in Japan in apparel and industrial fabrics, and in ropes and fishing nets.

There are some uses for water soluble fibres, and some poly(ethenol) fibre which is soluble is produced. Another water soluble fibre has been known since the 1940s, this is calcium alginate fibre which is dissolved by dilute alkali which converts it to the sodium salt. Alginic acid is obtained from seaweed and is closely related to cellulose:

Fibres are obtained by spinning sodium alginate solution into a slightly acidified solution of calcium chloride; calcium alginate is precipitated in filament form.

Water soluble fibres are used mainly as linking and support fibres, for example, socks may be knitted continuously in long strings with a short section of knitted alginate separating each. After closing the toes, the socks are separated by cutting, and dissolving out the alginate. Another use is as a base fabric for lace making, again being washed away finally to leave the finished, unsupported lace.

9

Polyalkene Fibres

Ethene and propene are the cheapest of monomers, produced in large quantity from mineral oil, and converted to polymers with wide and important use in the plastics industry.

I.C.I. chemists first polymerised ethene to high relative molecular mass in 1933, by using free radical initiators at above 100 MPa pressure.

$$nCH_2{=}CH_2 \longrightarrow (-CH_2-CH-)_n$$

Made in this way, the polymer melts at around 120°C, and has relative molecular mass about 20 000. Poly(ethene), more widely referred to as poly(ethylene) or polythene, is now a familiar, everyday plastics. Fibres were melt spun from it in the 1940s, but were too poor in physical properties for significant application. A major breakthrough came in the next decade with the discovery of new catalysts which not only polymerise ethene (and other alkenes) at low pressures, but also yield a product with much better physical properties.

Two low pressure catalytic processes were introduced. The Ziegler process uses organometallic catalysts which we briefly introduced in Chapter 2, and which we shall consider in more detail below. Ethene is polymerised at atmospheric pressure and below 100°C. The Philips (U.S.A.) process uses a chromium(III) oxide catalyst supported on silica and alumina, and ethene pressures of about 3 MPa at about 150°C.

Poly(ethene) may be made with a relative molecular mass of the order 10^5, and even up to 10^6 with Ziegler catalysts, it has a melting point of about 135°C, and a density of around 0.96 g cm^{-3}. Polymer made by the older high pressure process has a density of 0.915–0.935 g cm^{-3}. The difference in density gives some clue to the difference in melting points, for the higher density signifies a more closely packed molecular structure. The high density polymer is the more highly crystalline, and has a molecular structure which is most like the idealised linear molecule represented by the simple equation for polymerisation given above. But the molecules are not completely linear, they are branched by quite short side groups, mainly two and four carbon atoms in length. These branches arise via transfer reactions of the type noted in Chapter 2. The high density polymer has perhaps three such side groups per thousand carbon atoms, whilst the low density material polymerised by a free radical mechanism may have about twenty per thousand carbon atoms.

Fibres melt spun from high density poly(ethene) have some uses in industrial fabrics, upholstery fabrics, and in ropes and the like, but still have too low melting point for more general application in textiles. Poly(propene) fibres have been more successful, and we shall describe them in rather more detail. Total world production of all polyalkene fibres was about 250×10^6 kg in 1970.

Propene, $CH_3CH=CH_2$, is more difficult to polymerise than ethene using radical initiators. Development of the Ziegler catalysts by Professor Natta (Italy) in 1954 enabled him to polymerise propene to a highly crystalline solid melting at 175°C. This polymer has the regular isotactic structure (Chapter 2). In two dimensions this can be represented as:

$$
\begin{array}{c}
\; H \quad H \quad H \quad H \quad H \quad H \quad H \quad H \\
\; | \quad | \quad | \quad | \quad | \quad | \quad | \quad | \\
-C-C-C-C-C-C-C-C- \\
\; | \quad | \quad | \quad | \quad | \quad | \quad | \quad | \\
\; H \quad CH_3 \; H \quad CH_3 \; H \quad CH_3 \; H \quad CH_3
\end{array}
$$

An attempt at three dimensional representation is made in *Figure 9.1*, the structure is best appreciated if one builds a molecular model of a few units of the polymer.

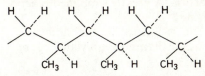

Figure 9.1   Isotactic poly(propene)

Propene is obtained by the thermal cracking of petroleum distillate fractions. The gas (bp -47°C) contains small quantities of several gaseous impurities, including CO, CO_2, O_2, H_2S, COS, ethyne, and other organics, which may all poison the catalyst. It is purified by fractionation, and sulphur compounds are adsorbed in soda-lime towers. Polymerisation grade material is better than 99.7% pure, with ethane as the main residual impurity.

Various Ziegler–Natta catalysts have been described in patents. A typical combination is titanium(III) chloride ($TiCl_3$) and diethyl-aluminium chloride, $(CH_3CH_2)_2AlCl$. These are highly reactive and hazardous chemicals, sensitive to moisture and oxygen. The aluminium alkyls are pyrophoric. Their preparation and handling on anything other than a laboratory scale poses many problems. The two are mixed in a dry liquid hydrocarbon (hexanes, heptanes, and homologues may be used) and react to form the active heterogeneous catalyst. Polymerisation is carried out at 25–100°C, and at pressures up to about 1.5 MPa. The isotactic polymer is insoluble in the hydrocarbon medium, so the product is a slurry of granular polymer. Any polymer molecules of low stereoregularity are soluble, and are therefore removed by the medium. Relative molecular mass may be controlled by a partial pressure of hydrogen gas, or by introducing small quantities of other compounds such as zinc alkyls. Control is necessary, otherwise the relative molecular mass becomes too high for subsequent melt fabrication—the melt viscosity is too great.

The temperature and pressure of polymerisation affect the product. At higher temperature and pressure the rate of polymerisation is greater, but the degree of steric perfection and the relative molecular mass are less than at the lower temperatures and pressures. The composition of the catalyst also plays a major part in determining steric regularity, rate, and degree of polymerisation.

Batch and continuous processes have been developed, and yields are said to be in excess of 10^3 kg of polymer per kg of catalyst. The catalyst in the slurry from the reactor is still active and is destroyed by hydrolysis with ethanol or water, the hydrogen chloride liberated is neutralised, and the polymer filtered off and dried.

The mechanism of polymerisation is not fully understood because of the variety and complexity of the catalysts, and the heterogeneous nature of the reaction. A greatly simplified reaction scheme is summarised in Chart 8.

The catalyst site is formed at the surface of the titanium(III) chloride crystal, where titanium is alkylated by the diethylaluminium chloride:

$$TiCl_x + Al(C_2H_5)_2Cl \longrightarrow TiCl_{x-1}(C_2H_5) + Al(C_2H_5)Cl_2$$

(Crystal surface)   (Crystal surface)

Propene is adsorbed on the surface, forming an unstable π-complex with titanium ions in favourably exposed positions which have vacant d-orbitals. A bond is then formed between the ethyl group at the catalytic site and the polarised propene molecule, with simultaneous interchange (Chart 8). A second propene molecule can then be adsorbed, and the insertion step repeated. In this way, by insertion of propene molecules between active site and polymer chain, the polymer molecule grows from the crystal surface. The steric regularity is determined by the spatial requirements of co-ordination of the propene molecules at the catalytic site.

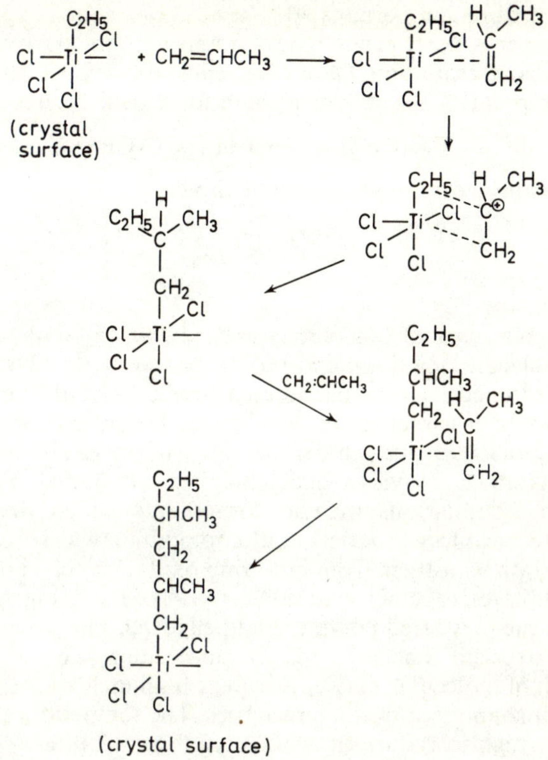

Chart 8   *Stereospecific polymerisation of propene—simplified mechanism*

Termination of polymerisation may be by reaction with alkyl halides,

$$Ti-P_n + RCl \longrightarrow Ti-Cl + RP_n$$
$$\text{(surface)} \qquad\qquad \text{(surface)}$$

here P_n denotes the polymer molecule of degree of polymerisation n. Or, by the adsorption and reaction of hydrogen at the active site,

$$Ti-P_n + H_2 \longrightarrow Ti-H + HP_n$$
$$\text{(surface)} \qquad\qquad \text{(surface)}$$

or a metal alkyl,

$$Ti-P_n + Zn(C_2H_5)_2 \longrightarrow Ti-(C_2H_5) + Zn(C_2H_5)P_n$$
$$\text{(surface)} \qquad\qquad\qquad \text{(surface)}$$

Poly(propene) suitable for melt spinning to fibres has a relative molecular mass about 2×10^5, with about 90% or more regular isotactic structure and perhaps 70% crystalline.

The polymer is fairly easily oxidised and is particularly sensitive to photoinitiated oxidation. The tertiary carbon atom, $—CH(CH_3)—$, is the weak point of the polymer chain for attack by free radicals,

$$R^{\bullet} + —CH(CH_3)— \longrightarrow RH + —\dot{C}(CH_3)—$$

Reaction of the radical with oxygen is rapid,

$$—\dot{C}(CH_3)— + O_2 \longrightarrow —\underset{\underset{OO^{\bullet}}{|}}{C}(CH_3)—$$

subsequent reactions of this peroxy radical lead to breakage of the polymer molecule (scission), and fall in relative molecular mass.

Radical initiated scission reactions cause a fall in molecular length when the polymer is melted for fabrication. To bring these changes in relative molecular mass under control, and to retard oxidation, antioxidants and ultra-violet light absorbers are added. The usual antioxidant formulations include compounds which decompose peroxides to harmless products, and compounds which react with radicals and prevent them from initiating oxidation reactions.

These additives, together with titanium(IV) oxide or pigments, are mixed with the powdered polymer and pelletised. The pellets are fed to screw extruders. The high relative molecular mass and broad distribution of molecular masses give very high melt viscosities, and melts have pronounced elastic properties. The temperature of spinning is very high to reduce the viscosity of the melt, usually about 280°C—a hundred degrees above the melting point.

The advantage of the cheap propene monomer has stimulated methods of making cheaper fibres. Coarse, ribbon-like fibres can be made by melt extruding poly(propene) film and slitting it by knives. A novel split-film technique was invented in 1954 by Rasmussen of Denmark. The extruded film is stretched to obtain parallel orientation of crystallites along the film. If a force is then applied, the film splits up into a sheet of thin fibres of rectangular cross-section. The sheet of fibres can be processed to coarse continuous filament yarns, or to staple, or can be converted directly to a 'non-woven' fabric. These products are used for carpet backing fabrics, outdoor carpets and surfaces, packaging and industrial fabrics, ropes, and twines.

Poly(propene) yarns are very difficult to dye because the hydrocarbon structure offers no ionic or polar groups for interaction with dye molecules: there is no attractive force to hold the dye molecule within the fibre. This great disadvantage, together with the relatively low melting point, has prevented any wide textile application of the fibre. The abrasion resistance is good, but not as high as that of nylon. Melt spun fibres can be drawn to high tenacity with a modulus intermediate between nylon and polyester, and with good resilience. They find use in ropes, cords, and nets (which float on water), and for carpet backings, upholstery fabrics and car seat covers, and for bristles. Chemical inertness (other than oxidative) permits use in industrial fabrics, filter cloths, laundry and dye bags.

The cheaper, coarse poly(propene) split- and slit-film products are penetrating markets previously held by jute. Fibres are also made by slitting and splitting poly(ethene) films, and have similar uses.

Because of the dyeing problem, polyalkene fibres are usually coloured by introducing pigments before melt extruding. Much scientific effort has been applied to finding practical solutions to the problem of dyeing these fibres, but with no general success. The usually fairly simple trick of copolymerising with small quantities of polar monomers is not possible with the Ziegler–Natta catalysts.

10

Elastomeric Fibres

Natural rubber threads find application in textiles wherever support or grip is required. Rubber has to be vulcanised before it is of use; this is a cross-linking process, covalently bonding together the polymeric molecules of natural rubber into a network, and preventing the relative molecular movements which bring about flow, or creep, under load.

Natural rubber has the repeating unit:

$$-CH_2-CH=C-CH_2-$$
$$\underset{\displaystyle CH_3}{\big|}$$

during the vulcanisation process covalent sulphide cross-links are introduced:

$$-CH-CH=C(CH_3)-CH_2- \quad .$$
$$\underset{\displaystyle S_x}{\big|}$$
$$-CH-CH=C(CH_3)-CH_2-$$

The chemistry is very complex and need not concern us here. Cross-linking prevents solution, for the molecules cannot separate, and so melt or solution spinning of vulcanised rubber filaments is impossible. To make filaments it is necessary to spin high quality natural rubber latex into a coagulating bath, and then vulcanise the threads by an after-treatment. Usually rubber is used as coarse support threads, which are made by slitting thin vulcanised sheet.

The production of synthetic rubber-like fibres which can be spun as fine filaments by accepted spinning technologies therefore offered an inviting target to the chemist. Of course, the immediate textile use of such fibres is very small compared with the huge markets of the synthetic fibres which we have already described, and so the development of elastic fibres has followed behind that of the others.

Synthesis of polymers with rubber-like properties presented a novel challenge to the fibres polymer chemist, he is normally concerned with hard, tough, materials of relatively low extensibility. The molecular architecture associated with high elasticity is quite the opposite of that for high melting tough solids. Elastic recovery from extension is brought about by thermal agitation producing a shortening of the molecules by rapid rotations about the covalent bonds of the polymer molecule. In the extended form the polymer molecule is in a state of higher potential energy. The chemist has to assemble molecules which are highly flexible at room temperatures, that is, which have a very low glass transition temperature. In addition, they should have little tendency to pack well together and to crystallise when oriented by stretching.

But to be of practical use as a fibre, the elastic substance must withstand under tension quite high temperatures associated with heat setting fabrics, hot water, and the repeated flexings caused by the movements of the wearer of the garment, without creep (i.e. irreversible extension), softening, or impairment of mechanical properties. Such requirements have to be met to some degree by all useful synthetic rubbers, and the usual way of obtaining adequate performance is to tie the molecules together to stop flow—that is, to cross-link them by some chemical reaction akin to vulcanisation!

One way out of this dilemma of conflicting properties is to spin a synthetic elastomer which can be easily cross-linked afterwards. Some commercial fibres are produced in this way; the filaments are solution spun and either cross-linked by chemical treatment in the coagulating bath, or by heat after wind up as yarn. But such processes, involving treatment of miles of filaments, are difficult to control to bring about uniform and reproducible properties, and may be expensive in terms of processing time.

An alternative approach is to introduce physical cross-links rather than the covalent chemical cross-links, by building into the rubbery molecule short lengths of molecular structure which are high melting: these polymers are classed as block polymers.

A block polymer molecule is built up by end to end combination of sequences of different monomers. Thus, the random copolymer of monomers A and B can be represented as:

—ABBABAAABBABAABBBBABAABBB—

the block polymer of monomers A and B has the more ordered structure:

—AAAAAAAAAA—BBBBB—AAAAAA—BBBBBBB—

The lengths of the sequences of A-units and of B-units will not usually be constant, but there will be some statistical distribution of lengths. By changing the average lengths of the blocks of A or B, the proportions of the two components, and the order in which the blocks are assembled (a three block sandwich A_x—B_y—A_z has different properties from a sandwich B_p—A_q—B_r) a very wide range of polymeric materials and of properties can be obtained. The separate block components contribute to the whole something of their individual polymer properties so that interesting combinations of properties can be achieved leading to quite unique materials.

In the elastic block polymers long segments of the molecules consist of flexible low glass transition polymer, and are separated by much shorter segments of high glass transition, rigid, perhaps crystallisable, polymer. In very simple terms, associations or clusters of the latter segments by their internal stiffness and inter-molecular forces provide the rigid physical cross-link which holds the molecules together in the solid state and prevent creep: the long flexible segments impart the rubber-like elasticity. A diagrammatic representation of the structure is given in *Figure 10.1*. The block

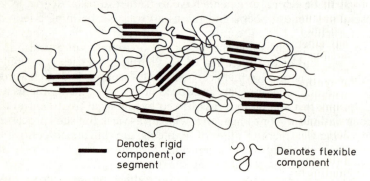

Denotes rigid component, or segment

Denotes flexible component

Figure 10.1   Formation of 'spandex' elastomeric fibres. Diagrammatic block polymer concept. Associations or clusters of rigid segments hold assembly of molecules together. Flexible interconnections provide long range elasticity

polymer is soluble in suitable liquids, and at some temperature the associations are disrupted and it will soften and flow as a viscous liquid (there is no sharp melting point). Thus, fabrication to fibres, and other shapes, is possible by extrusion of the melt or of a solution.

The design of block polymers to provide specific physical proper-ties is an exciting scientific advance, which opens up wider possibilities for new materials, but they are far more complex in chemistry and in technology than the copolymers and polymers in established use, and are correspondingly more expensive to pro-duce. Thus, future development is likely to be restricted to specialised materials—such as extrudable rubbers.

Du Pont introduced 'Lycra' the first commercial synthetic elastic fibre in 1958, and several others now on the market are essentially

Formation of "spandex" elastomeric fibres.
Example to illustrate chemistry used:

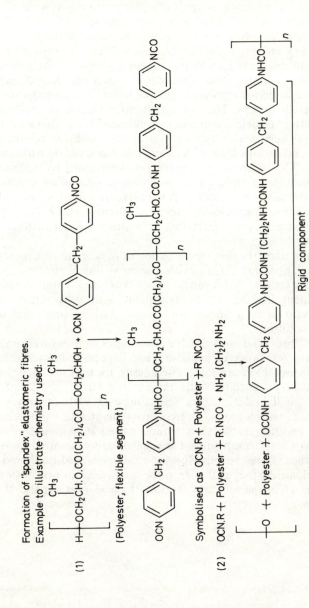

Chart 9   Formation of 'spandex' elastomeric fibres

similar in chemical type and based on polyurethane chemistry. The generic name of 'spandex' is applied to designate elastomeric fibres consisting of at least 85% of a block polyurethane.

An aliphatic copolyester or polyether is taken as the low glass transition, low melting point, flexible segment, and is prepared with hydroxyl end-groups (Chart 9). Its number average relative molecular mass may be about 2000. The hydroxyl ends are reacted with a diisocyanate so that the polyester (or polyether) is now end-capped with isocyanate groups. The isocyanate group in organic chemistry is very reactive. The rigid segment is introduced by linking together the flexible segments with a diol or a diamine which reacts with the isocyanate end-groups (and with excess free diisocyanate which will be present). The choice of the reagents is determined by balancing ultimate fibre properties, price, and considerations of chemical reactivity. It is fairly wide, and each manufacturer has his own polymer composition (undisclosed, except in patents), and polymerisation and spinning processes: this is another unique feature of elastomeric fibres.

Depending upon the polymer, solution or melt spinning processes may be used. Spandex fibres can also be cross-linked after spinning by using an excess of diisocyanate which is reacted within the fibre by subsequent heating, or by treatment with other chemicals. 'Lycra' is said to be a linear, non-cross-linked polymer, and dry spun from solution.

Spandex fibres offer considerable advantages over natural rubber in toughness, resistance to oxidation, and especially in availability as fine filaments. They are rarely used alone, for they are expensive, and the rubbery feel they would impart to a fabric would not be pleasant. Usually elastic fibres are combined with other textile fibres, perhaps with a bulked stretch yarn (Chapter 12), or as a core yarn with another textile yarn wound round it. Alternatively, a spandex yarn may be introduced at regular intervals into a knitted fabric. As these fibres become further established and improved in quality and variety they will offer to the textile and fashion designer new ranges of properties and fabrics which should expand their application.

11

Inorganic Fibres and Carbon Fibres

Fibres prepared from inorganic materials have not found wide use in conventional textile applications, but present a range of physical properties which are quite distinct from those of the organic fibres which we have discussed. The most important by far in terms of quantity of production and versatility is glass fibre.

Glass fibres were described by Hooke in 1665 and were known to the Ancient Egyptians. The first glass fabric was made in 1893, and a dress was worn by an actress that year at the Colombian Exposition in Chicago; it is said the event aroused a great deal of public curiosity, no doubt prompted by the familiar transparency of sheet glass. Modern industry dates from the 1930s when the Owens–Corning Fibreglass Corporation was set up in the U.S.A. Fibre was used initially for thermal and electrical insulation, and for filters, but subsequently has been developed also for fire-proof curtains, protective clothing and other fabrics, for plastics reinforcement, and for rubber reinforcement including tyres. Its use in tyres brings glass fibre into an area of keen competition with the man-made organic textile fibres.

Molten glass is allowed to flow through 'spinneret holes' in the bottom of a furnace and filaments are then drawn away at very high speeds from the extruding high viscosity melt (speeds in excess of 3000 m min^{-1} may be used), and wound up as continuous filament. Glass staple fibre 20–40 cm long is obtained by using high velocity air jets to draw fibres from the orifices. (Shorter staple fibre may be obtained by further cutting.)

The mechanical properties of fine glass fibres are quite remarkable, the tensile strength may be $35\,000 \text{ kgf cm}^{-2}$ (about 15 g dtex^{-1}); by comparison bulk glass has a tensile strength of only 1000 kgf cm^{-2}. (The high tensile strength is still only about a quarter of the theoretical strength.) The relatively high tensile strength of glass filaments seems to be due to the absence of fine cracks at the surface, for if the surface is damaged by handling the strength falls, and due also to the rapid cooling of the filament which sets up differences in stress between the surface and the interior. The self coefficient of friction of glass is high, and its abrasion resistance is poor, so to protect the surface a lubricant is applied at spinning before the fibres are wound up. In addition to surface protection, a surface treatment to promote bonding to specific resins is needed for glass fibre for plastics reinforcement.

The elasticity is perfect Hookean, without a yield point, quite different from the behaviour of all other fibres we have considered, but the extension to break is only about 4%. The Young's modulus is about 300 g dtex^{-1} (7×10^5 kgf cm^{-2}), which is of the order of ten times greater than the established textile organic polymer fibres. The high speed draw-down of glass filaments at spinning gives very fine filament diameters, so the fibres are flexible and soft despite the high tensile modulus and low extensibility.

The physical properties of glass, together with its zero moisture regain, give rise to fabrics which have excellent dimensional stability. Additionally, glass is not degraded by exposure to sunlight, has very high resistance to chemicals, can be used up to about 500°C, and will not burn.

The uses of glass fibre for insulation, for filtration of hot gases or reactive chemical solutions, and for plastics reinforcement, and so on, follow naturally from the properties. The poor abrasion resistance, low extensibility, and problems of dyeing glass fabric seriously restrict its use as a textile fibre in apparel fabrics.

The strength, high modulus, and elasticity of glass fibre make it excellent for the reinforcement of plastics. Commercial plastics are not very strong or stiff in comparison with constructional metals, but the introduction of reinforcing fibre produces a composite material of greatly improved mechanical properties. The fibres carry the load, and the matrix polymer prevents the fibres abrading and maintains the structural shape of the object. Composites of this type promise to be even more important in the future as structural materials, and so the reinforcing fibres present an area of current scientific and technological activity, although markets for the more exotic (expensive!) fibres are yet very limited. The quantities of cheaper fibres used throughout the world to reinforce plastics are already very large, for several millions of tonnes of such materials are produced each year: glass, natural asbestos, cellulosic, and synthetic organic fibres are all employed. Wider use and application of reinforced plastics is unlikely until means for automating their fabrication are further developed.

The types and applications of glass fibre reinforced plastics are extremely varied. Several polymers are commonly used as matrix material, but mainly polyester resins. Applications are as diverse as boats, car bodies, aircraft and rocket components, building panels, electrical components, gears and bearings. One big advantage is that large and complex structures can be assembled quite easily, although by hand. The glass reinforcement is supplied either as staple, continuous filament, fibre mats, or woven fabric depending upon the performance requirements of the end product.

Glass fibres for rubber reinforcement need special care to preserve the full tensile strength of the spun filaments and a method has been developed for impregnating the fibres so that they do not abrade each other but adhere firmly to the rubber. Glass reinforced rubber drive belts have excellent flex life and dimensional stability, and there is active development in the valuable tyre market where the high modulus of glass fibre is suited to radial tyres.

Several other inorganic fibres have been developed recently, primarily to meet the demands of the modern aircraft industry and the development of rocket motors and space vehicles—areas in which a very expensive specialised material can find a market. The need here is for low density, very stiff and strong fibres which can be used at high temperatures in metal composites as well as in plastics. Some extremely ingenious work has gone into the development of these materials and probably the longer term future exploitation may lead to cheaper products and wider use.

Aluminium silicate fibres can be spun much as glass fibres, and are used for high temperature insulation, filters, and so on. An extension of the viscose spinning process has been developed to produce some other ceramic fibres: suitable metal oxides are dissolved in the alkaline viscose solution and a fibre spun. The cellulose is then carbonised by heating to a high temperature, followed by a further high temperature oxidation and sintering treatment, to burn off the carbon and yield the ceramic fibre. The cellulose acts as a carrier to enable fabrication of the inorganic material.

Boron fibres are made by deposition of boron vapour onto a fine tungsten wire substrate, and are used in metal composites for the aerospace programme in the U.S.A.

Very strong fibres have been obtained by growing single crystal whiskers of metals, and of several inorganic compounds, from the gaseous state. These whiskers may vary from a few millimetres to several centimetres in length, and closely approach the theoretical strength of the substance. However, so far there are no realistic processes for their production on an industrial scale.

Metallic fibres are produced (and indeed have the claim to the oldest man-made fibres, since threads of gold and silver were used in ancient times to decorate clothing), and still used as decorative threads in textiles. Fibres are obtained by the drawing down of wire, and by slicing metal foil which is usually backed or laminated with transparent plastic foil. For decorative textile applications aluminium foil is used, and the plastic foil may be coloured for effect.

Steel wires are widely used as reinforcing cords in tyres, and recent developments in the melt spinning of finer steel 'fibres' suggest that the future challenge to organic fibre tyre cords will be even greater. They also have special high temperature applications.

The ultimate in low density, stiff, and strong fibres are the newly developed carbon fibres. Carbon fibres made by pyrolysis of cellulosic fibres have long been known, but have no remarkable mechanical properties. The new carbon fibres in contrast have been specifically produced for use in high performance composite materials. The principle for attaining high strength and modulus is nothing new to the man-made fibre industry; it is to obtain high orientation of the graphite crystallites along the length of the fibre. American workers achieved this by further stretching pyrolysed viscose rayon fibres at over 2000°C, which is quite an extension to conventional textile hot stretching techniques! British scientists, W. Johnson, L. N. Phillips and W. Watt, at the Royal Aircraft Establishment, Farnborough, began work on carbon fibres from poly(cyanoethene) fibres in 1963. Supported by Rolls-Royce, Courtaulds, and Morganite, these fibres were on sale by 1968, which shows that we can exploit inventions at speed provided that they are recognised early and properly backed.

Poly(cyanoethene) fibres, spun as described in Chapter 8 blacken, without melting, on heating in air at below 300°C. The linear molecules change to a fused ring structure,

Reaction with oxygen removes some of the hydrogen atoms and leads to build-up of layers of fused rings. Freely supported fibres shrink, but if they are held fixed to length during this oxidation stage the shrinkage forces produce a tension in the fibres which maintains and improves the initial spun orientation. This is a key step in obtaining good mechanical properties, and a very neat and simple solution to the problem of maintaining orientation.

Carbonisation is completed by heating in inert atmospheres at above 1000°C, when hydrogen and nitrogen atoms are expelled to leave a fibre composed of oriented fibrils of hexagonal rings of carbon atoms. To obtain complete, crystalline, graphitisation, and the high modulus, a final heat treatment of 2500°C or higher is applied. There is no need to hold to length here. The stress–strain curve is linear, as for glass fibre, with a small extension to break. Modulus and tensile strength depend upon the shrinkage forces experienced, and upon temperatures and times of final heat treatment, a modulus of about 3.5×10^6 kgf cm^{-2} (340 GN m^{-2}) can be attained.

Carbon fibres are black, but surprisingly silky in appearance; their future lies in the more expensive composites designed for hard work at extremes of temperature and for engineering structures; for example, aircraft and spacecraft structures and components, fan blades for aero-engines, rotor blades, bearings, and pressure vessels. Electric power cables of aluminium and carbon fibre are being studied, offering high conductivity and the strength of high tensile steel. Whilst glass fibre will always be much cheaper and must retain the much larger bulk market, carbon fibre can be used to stiffen glass fibre composites, and such combinations promise to extend the application of glass fibre reinforced plastics to larger structures and to new outlets.

However, the impact of high modulus aromatic polyamide fibres on the structural composites market remains to be seen. Whilst carbon, boron and other expensive fibres should retain their specialised high temperature duties, the new polyamide fibres on price alone must compete favourably for the less exacting roles.

12

Yarns and Fabrics

The spinning technologies described in Chapter 3 produce continuous filaments which are drawn or stretched to develop anisotropic fibre properties. Some of these yarns, twisted and wound up on cylinders of cardboard, plastics, or metal, are the end product for the fibre producer, and are sold to his customers who will process them into the myriad of goods for which they are destined. However, there are other stages in the production of yarns which may be carried out by the fibre producer, or by customers, and in addition there are new developments in the direct production of fabrics from the spinning machine.

Twisting to give filament cohesion is a traditional and relatively slow process, developed for staple fibre spinning. Alternative, faster methods applicable to continuous filaments can be used for some yarns.

If a bundle of parallel continuous filaments is passed through a controlled air jet, they can be made to oscillate, and to intermingle and interlace to form a cohesive yarn without twist. By feeding two bundles of different filaments together into the air jet a blend into one yarn is obtained. The filaments may be different in diameter or cross-section, or may be of two completely different polymers, for example nylon and rayon.

A more turbulent air jet produces loops in the filaments as well as entanglements. Thus, a 'bulked' yarn is produced from the 'flat' continuous filament, and the loops cannot be straightened out by pulling the yarn because of the entangled structure. This air jet method of giving bulk, or texture, to a yarn is applicable to all continuous filaments. The yarns are especially useful for weaving, because the bulk is not pulled out when the yarns are under tension in the loom.

There are several other methods in use for bulking continuous filament yarns, and the commercial value of the products has stimulated considerable activity in devising better and faster processes. Each method produces different yarn structures and appearance, and some synthetic fibres respond better than others. Examination of assorted discarded garments and scraps of fabric with a lens will reveal something of the variety of yarn textures—and fabric structures.

If thermoplastic yarns are deformed when hot and cooled before release, the deformation is set in position (Chapter 2). This principle is the basis of several bulking methods.

A saw-tooth crimp is made by feeding heated yarns into the teeth of intermeshing gear wheels, or fluted rollers, and by compressing yarns into a heated confined space. In the latter method, the yarn is pushed into a hot box or tube (the size depending upon tex of the yarn) by feed rollers, and cools as it is continuously withdrawn at a slower rate from the exit end of the device. The process is descriptively known as 'stuffer-box' crimping. The crimp will pull out straight when the yarn is under tension, but gear and stuffer-box crimping methods are capable of dealing with heavy carpet yarns, and with heavy tows which are to be cut to staple (see below).

If a heated yarn is passed round a (blunt!) knife edge the filaments are distorted by bending, the inside in contact with the edge is compressed, and the outside extended. The asymmetric stresses set up across the filaments are relieved, when the yarn is relaxed, by coiling each filament into a helix. The yarn is thus composed of miniature coiled springs and demonstrates high extensibility (although the restoring force is low). It is a stretch yarn. Stretch yarns exhibited a property novel to the textile industry when they were first introduced, and contributed greatly to the growth in popularity of knitted fabrics with application in sportswear, sweaters, underwear, stretch tights and stockings, and stretch covers. The open construction of a knitted fabric allows full use of the stretch properties of the yarn.

The most developed and commonly used methods of texturing continuous filaments to make stretch textile yarns are false twist devices, which can be run at very high speeds. If a yarn is twisted to a high degree (up to 40 turns per cm) and heated to set it in the deformed state, and cooled, then untwisting the yarn again will not restore the initial state. The untwisted yarn will be coiled, unstable, and will try to twist up once more if released. A stable bulked yarn can be obtained by combining it with another yarn which has been twisted in the opposite direction so that the restoring forces are balanced. Twisting followed by untwisting is clearly a slow and expensive way to make a bulked yarn. The false twist process works on the principle of a length of yarn held at both ends and twisted in the middle, equal and opposite twist is put in at each side—hence the term false twist. If a continuous process is set up with the length of yarn running over some device for twisting it, for example in contact with the rim of a rotating disc, then twist is inserted in the length of yarn approaching

the disc, and taken out in the length moving away from the disc.
A heater on the yarn feed side will set the twist on that side, and so
there is a continuous twist-set-untwist sequence in one compact unit.
The twisting device can be run at high speed, and hence yarn can be
processed at speed.

Other methods in use include the twisting of two yarns together,
heating, and then separating the two; knitting a yarn, heating, and
pulling out yarn from the knitted structure—the stitch deformations
are set in position. (A similar effect on a large scale is observed by
pulling back a knitted woollen garment.) These methods are carried
out continuously, of course.

There may be too much stretch in a yarn, and extensibility can be
controlled (limited) by the usual trick of heating, or steaming, in a
partially extended state to allow the structure to relax itself and
stabilise in the new position. A heat relaxation, or setting, is desirable
for dimensional stability of fabrics made from the yarns, so that they
do not loose stretch during dyeing, and washing.

Bulked yarns are also obtained from bicomponent fibres. Different
strains are set up in the two components when the filament is drawn.
When the drawn filament is subsequently relaxed by heat, or by
boiling in water, the asymmetric shrinkage forces cause it to coil up
into a helix. The mechanism is similar to that of bulking over a knife
edge.

The forces involved are small, and the structures of the yarn
and of the fabric restrict the full development of a helix. The filaments
appear as extended helices, with a wavy crimp. If one of the two components of the filament has a low glass transition temperature, and
is non-crystalline, for example an elastomer (Chapter 10), the
filament will coil up spontaneously as soon as the drawing tension is
taken off.

Differential swelling and shrinkage forces between the two components when they are immersed in water can cause bulking of yarns.
Nature sets the example with wool. Asymmetric viscose rayon fibres
can be produced by special bath conditions, or two different dopes
can be spun together as a bicomponent. One side of the asymmetric
structure swells more than the other in water. Bicomponent acrylic
fibres were introduced in 1959 (Du Pont). Composed of two different
acrylic copolymers, differential shrinkage and crimp occurs on
relaxation in boiling water. This crimp is reversible (like wool) on
wetting and drying the fibre. The higher shrinkage component swells
the most in water and so tends to straighten out the crimp, which
is recovered on drying. This fibre movement, without felting,
is of advantage in restoring the appearance of a garment at
laundering.

STAPLE FIBRE AND YARN

The production of staple fibre begins with the spinning or collection together of many thousands of continuous filaments from the spinning machine. This bundle of parallel filaments is known as tow. It may be hot drawn or stretched, crimped, and fed continuously through rotating cutters to be chopped to short lengths. Multiple carding and combing steps are necessary to disentangle wool and cotton fibres and lay them in some sort of parallel array prior to spinning. Man-made staple fibres are treated in similar fashion, but it is also possible to avoid these processes by maintaining the parallel order already present in the continuous filament tow.

In one method, the tow is opened out to a sheet of filaments, drawn or stretched between heated rollers, and then cut continuously by a revolving knife. The cut must be diagonal across the sheet so that the cut ends overlap, otherwise the cut tow will have no cohesion. Cutting compresses the ends of fibres together, and so is followed by separating stages between fluted rollers, and by a thinning down of the sheet of fibres (for example by passing through a set of rollers at progressively increasing speeds). The thin sheet of short, parallel fibres is then rolled up into a loose twistless rope called sliver. Crimping can be carried out on the drawn tow or on the sliver, using stuffer-box or fluted roller methods.

An alternative method of making staple is to stretch the tow between two sets of rotating rollers until the filaments break. A filament will break at its weakest point between the rollers and so the broken fibres will be of variable length and overstretched. These defects are avoided by bringing the taut stretched filaments into contact with sharp edged fluted rollers. The filaments then snap at the sharp edge without being overstretched, and staple length is controlled by the distance between breaking edge and the nip of the stretching rollers. A sliver is formed and crimped, as above. This stretch-breaking method of making staple is used especially for polyester, acrylic, and polyamide fibres. The staple may be relaxed in steam to allow a shrinkage recovery from the stretch.

At suitable points during the production of sliver by either cutting or stretch-breaking tow, a tow or sliver of other man-made fibre, or a natural fibre sliver, can be fed in and combined to make a blend of staple fibres in the final sliver. Blends of synthetic and natural fibre are also prepared by mixing staple at the carding stage (see below), so that an intimate mixture is obtained. Blending of different diameter staple fibres of the same material, or fibres of different shrinkages, is also carried out to obtain a variety of yarns with different characteristics. The length to which staple fibre is cut is determined by the blending and spinning system for which it is destined.

Carding is the process of separating the entangled staple fibres. It is carried out by teasing out the fibrous mass between two sets of inclined stiff wire bristles. The carded fibres come from the machine as a thin sheet or web of fibres which is taken off as a sliver. Combing is carried out to introduce some parallel alignment to the fibres and to take out short fibres. Wool is combed for worsted spinning, and some long staple cotton is combed after carding. The slivers from carding and combing machines are rather irregular in fibre density, so several are combined and drawn out between pairs of rollers rotating at successively increasing speeds to make a single, more uniform sliver.

The spinning of yarn from sliver is carried out by a combination of reducing the diameter by drawing out between rollers and introducing twist to provide cohesion of the overlapping fibres. Crimp gives greater fibre cohesion.

NON-WOVEN FABRICS

Fabrics are traditionally made by either weaving or knitting. Weaving is the more obvious (and more ancient) process of interlacing a yarn (the weft) carried back and forth in a shuttle across a sheet of parallel yarns (the warp). Knitting is the process of continuously interlooping a yarn using a system of needles. Simple woven and knitted structures are illustrated diagrammatically in *Figure 12.1*.

Both processes are very highly developed and sophisticated, capable of producing a very wide range of excellent fabrics. Modern machines, especially knitting machines, operate at high speeds (0.2 m min^{-1} of fabric).

However, no matter what the machine speed, yarns for weaving and knitting must be made by the lengthy and complex methods outlined in this book. Is it possible to go directly from spinning machine to fabric? A great deal of technical effort throughout the world is being applied to this possibility.

Fabrics are being made by distributing the man-made continuous filaments on to a conveyor belt which in effect replaces the wind up device of the spinning machine. An air jet system is commonly used to accelerate the filaments and direct them on to the moving belt. The defect of static electrification can be put to good use to separate the filaments in flight and to position them on the belt. The deposited filaments form a loose web on the conveyor, with little or no cohesion, and must be bonded together at fibre cross-over points to

Warp yarns

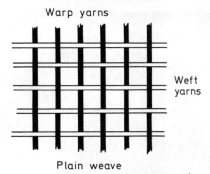

Weft
yarns

Plain weave

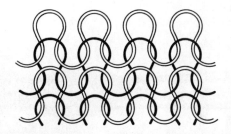

Simple weft knitting

Figure 12.1

form a continuous network.

Bonding can be brought about by applying an adhesive which is dried and hardened by heat. The adhesive is better applied in discrete spots rather than over the whole area. Total area bonding gives a stronger but very stiff and harsh structure, whereas spot bonds allow some free relative movement of fibres at those cross-over points which are not bonded and the structure is more flexible, but weaker. Thus, a spot bonded web has properties of flexibility and softness more like a conventional fabric.

The adhesive can be supplied by blending into the spun web a small proportion of a lower melting thermoplastic fibre. The web is then bonded by heating above the softening point of this fibre, but below the melting temperature of the principal fibre of the web. The low melting fibre fuses and, on cooling, bonds together random sections of web.

I.C.I. Fibres have devised a melt bonding method which makes use of bicomponent fibres. One component of the bicomponent fibre is lower melting and acts as the glue! This may form the sheath

Figure 12.2   A photomicrograph to show the bond structure of I.C.I. Fibres melded fabrics

of a core/sheath fibre, or the two may be spun side by side (Chapter 3). Two convenient polymers for use are, for example, 6.6 nylon (mp 265°C) and 6 nylon (mp 220°C). When the web is heated bonds of 6 nylon are formed where bicomponent fibres touch each other. There are no 'blobs' of surplus fused polymer. The fabric structure is illustrated in *Figure 12.2*.

The proportion of bicomponent fibre can be varied to control the density of bonds in the structure. Bonding can be overall by a general heating of the whole web, or at discrete points and in patterns, for example by passing beneath a heated engraved roller. The method has the advantage of providing a latent adhesive which is available for use at any part of the web, but which otherwise is an integral part of the fibrous lattice and makes a full contribution to the physical properties. The description 'melded fabric' is applied to fabrics bonded by the I.C.I. method.

The above developments of directly spun continuous filament fabrics from polyamides, polyesters, and polyalkenes, are quite recent, and the techniques are only in the early stages of commercial use. Many new ideas, and more sophisticated fabrics can be expected. Du Pont introduced spun-bonded polyalkene fabrics in 1964; I.C.I.'s melded fabrics were then being made in the Pontypool Research Department. However, the production of non-woven fabrics from staple fibres predates those from continuous filament by a few thousand years!

The preparation of woollen felts by squeezing and beating wool in warm water was mentioned in Chapter 4. Paper is made from short cellulose fibres (in wood pulp, etc.) which are deposited from aqueous suspension on to a porous support, and dried. The fibres are drawn together by the surface tension of the water film, and then hydrogen bonding at the surfaces gives excellent cohesion.

The modern production of non-wovens from man-made staple fibres follows essentially similar steps to the continuous filament processes, i.e. the formation of a web of fibres followed by a bonding stage. Webs are made by carding techniques. The carded web is fragile, with little strength across the width, for the wires of the card lay the fibres more nearly parallel, and it is usual to lay webs at an angle, one on top of the other, to obtain strength and build up the desired weight. To prevent delamination of the layered structure it may be punched through by barbed needles which carry fibres through the web from top to bottom. This is a commonly used method for consolidating any loose fibrous web.

A random array of fibres is obtained by using an air blast to pick up and separate staple and deposit it upon a collecting screen of perforated metal. Alternatively, the technique of the paper-maker is borrowed, and fibres deposited from a slurry in water. Handled in these latter ways, the fibres suffer no mechanical damage, and webs from paper thin to several centimetres thick can be deposited.

A felt can be made from synthetic fibres by needle punching a heavy web to a highly entangled structure which is further consolidated by fibre shrinkage in hot water or air. Such felts have formed the basis of some synthetic leathers; the dense fibrous felt is filled with a porous polyurethane rubber and given a surface coating of tough polyurethane.

Bonding of staple webs is carried out by the same methods as used for continuous filament webs. Staple offers its usual greater flexibility in permitting easier blending of different fibres, and allows a wider range of fabric thicknesses. But bonded staple fabrics are generally not as strong as continuous filament fabrics of the same density. They are used for the thicker, denser fabrics, such as floor coverings, and upholstery. I.C.I.'s melded staple process and 'Cambrelle' fibre is being used to produce hard wearing, attractive carpets (the 'Endura' range of Gilt Edge Carpets).

So far, most non-woven fabrics have been used for the cheaper, non-aesthetic textile outlets such as interlinings for suits, packaging and filter cloths. The stronger continuous filament products are additionally used for carpet backings and as reinforcing fabrics to be coated with plastics for use as protective covers. The lower cost of making the fabrics now enables the more expensive melt spun synthetic fibres to compete with rayon, cotton, and jute in these markets.

Now the greater challenge is being taken up of making lighter weight fabrics, and fabrics with drape and suitable for curtains, and even clothing. Lighter weight fabrics are being used for the so-called disposable articles—paper handkerchiefs and serviettes are commonplace examples. Disposable underwear is available, and other uses, for example hospital sheets and linen where laundering may be less desirable than destruction, are likely to grow. But, the main goal is to make fabrics with drape, and good appearance and handle which can be used in the home in fair comparison with traditional knitted and woven fabrics. The core of the problem is to make a non-woven with sufficient fibre mobility to give the properties of drape, and yet to provide strength against tearing and prevent the loss of fibre by abrasion at the surface. The structures of spun yarns and of woven and knitted fabrics are ideally suited to do these things!

Bonding by adhesives is one approach to the production of fabrics. An alternative route is by a mechanical interlocking of the web, for example by development and refinement of the needle punching technique so that the fibres are pushed and interlaced into some sort of ordered structure. Blankets are made in the U.S.A. by this method. A number of machines have been developed in East Germany, Czechoslovakia, and Russia, which produce a variety of stitch-bonded fabrics. In one fabric construction a web of staple fibres is bonded by knitting a stitching yarn through the web. Alternatively, warp and weft yarns are laid across each other and then stitched together by a stitching thread. These fabrics can be made light in weight, with reasonable drape, and can be used for curtains and apparel. The heavier stitch-bonded webs are used for blankets, upholstery, and heavier apparel fabrics. Fabrics can be made at speeds of from one to five metres per minute.

Non-woven fabrics of all kinds, adhesive bonded, melded, spun-bonded, stitch-bonded, and others to be invented, are going to have a great impact upon the future of the textile industry, and upon the style of fabrics and clothes which we shall wear. They will certainly not displace the traditional woven and knitted fabrics and garments entirely, but rather will provide an additional choice of less expensive fabrics, and of new fabrics. The whole pattern of the industry may well be changed to some extent, for the fibre producer—the chemical industry—can now make fabrics directly from his fibres.

13

The Future

This book is intended for those who will help make the future, and any attempt at prediction is undoubtedly presumptuous and is sure to be incorrect, but some trends of the present day textile industry can be extrapolated within the remaining years of the century. The textile crafts are ancient, and mechanisation and industrialisation of these crafts dates back about 200 years, but the effective development of the man-made fibres has taken place within about 70 years. Without doubt this initial phase of intensive invention and development of new fibres is drawing to a close. The immediate future will see continued and increasing production and use of the existing man-made fibres throughout the world, as populations and as living standards rise. Man-made fibre industries are growing in all countries, although the research and technological innovation is still coming mainly from the U.S.A., Japan, and Europe. The textile industry as a whole, in terms of volume of fibres handled, is going to be increasingly centred in the highly populated countries of the East.

Some of the newer fibres we have mentioned will flourish; this seems especially true of the new very high modulus fibres (organic and inorganic) which fill a gap and are creating greater opportunity for tyre cords, and particularly for reinforced plastics and fibre-reinforced concrete as future constructional materials. Some of these materials may one day replace timber in many applications, appropriately enough, for wood is a natural fibre-reinforced composite. The fire hazard of plastics currently limits use for many constructional purposes. New materials may arise by crystallising existing polymers in extended chain form (Chapter 2).

Individual success or failure of the new fibres will depend upon ultimate price per kilogram for a given unit of performance. The market for glass fibres will certainly grow in the composites area.

However, it is unlikely that any of the major established textile fibres will decline to the point of extinction within this century.

Within our present overall technological concepts and development, fibrous materials (natural or synthetic) exist covering a sufficiently broad range of physical and chemical properties to satisfy our basic needs. The obvious deficiencies in the properties of the synthetic fibres remain: we would like improved antistatic properties, and additives to give better protection against photo-oxidation. Currently there is great activity in a search for agents to provide non-flammable fibres. We may expect continuing progress in the processes of colouring and dyeing fibres, and in the technology of finishing yarns and fabrics.

Further physical modifications, and improvements in manufacture and processing the existing fibres will continue, with the most important developments in the methods of making textured yarns. The improved textured yarns will displace some staple fibre, and lead to a greater variety of fabrics.

It is in the ways of making fabrics that the greatest changes are taking place now, as described in Chapter 12, and we may expect this to be an area in which the most radical innovation will continue. Following a phase of invention of fibres, it seems logical evolution that there should be greatest activity in yarn processing, in fabric making, and then in new ways of using fabrics and of making garments and articles. Keeping pace with the new fabrication methods will be different ways of colouring and patterning, and a new challenge and opportunity to the textile and fashion designer.

The position of the natural fibres is more difficult to assess. One view is that their use will decline rapidly, but the production of vegetable fibres can be a vital source of income to a poor country and a native industry not easily replaced, whilst vast tracts of Australia seem well fitted for sheep farming for a long time to come.

Eventually pressures for land, and the cost of producing wool especially, will probably cause a decline, but it is perhaps likely that the level of production on a world scale will stabilise at about the present quantities for some considerable time. Wool has suffered most from the impact of the new fibres and its high price makes it most vulnerable for the future.

One strength of the natural fibres, cotton and wool, is that they are not adequately matched in all properties by polynosic rayons and synthetic fibres. This is especially important of moisture regain and comfort in wear.

The longer term scientific development of the textile industry is of course dependent upon the social, economic, and political future of the world, and upon future scientific discoveries and knowledge at which we can hardly guess. It is possible to anticipate some of the problems which will arise quite soon. Where shall we find the chemical intermediates to produce vast quantities of synthetic fibres and plastics? How shall we responsibly dispose of the used and waste fibres (which are not biodegradable)? Coal will provide a source of organic chemicals long after the known oil reserves are exhausted, but no doubt we will find it desirable, even essential, to process scrap organic fibres to recover organic carbon. This in turn may lead to the invention of new fibres, to replace our existing ones, which will be biodegradable and probably will be protein-like or synthetic cellulosic. These new materials could be made by chemical processes derived from biochemistry and more akin to natural processes, using new organic catalysts working on the principles of enzymes, far removed from the crude, energy-wasteful, high temperature and pressure chemistry of today.

Development of this new chemistry will of itself make available new commercial intermediates and lead to new fibres: the second phase of fibre invention perhaps.

As far as the availability of cheap, unlimited raw materials goes, inorganic fibres based on silicates offer the ultimate attraction. But at present there seems no way of making versatile textile fibres from inorganic polymers. This may be another point of scientific breakthrough.

Speculations of fibres with novel properties which may be invented in the future have ranged over new electrically conducting materials which will provide heat and light, or find application in solid state electronic devices, and fibres which are sensitive to humidity, heat, or to light, and will close or open up a fabric structure in response to changes in ambient conditions. Such unique fibres may be possible, and be made, but the large markets must always remain open to the adaptable, versatile, general purpose textile fibre.

Will fibres be needed in the future, or will garments and fabrics be made from non-fibrous, leather-like, porous sheet?

Already some considerable steps have been taken in making such materials, for example polyurethane foams, and notably the porous polyurethane 'Porvair' for shoe uppers. But at present the wider penetration of these and similar non-fibrous poromerics into the general textile field seems remote, although they will have limited specialised use for example as linings, upholstery, and in outerwear as dictated by fashion. General applicability as textiles is restricted by the need (at present) to make fairly dense or thick structures to obtain adequate physical properties, by the difficulty of matching the flexibility and drape of a fabric, and by lack of versatility. For here are the real fundamentals of fibres and of fabrics; the uniquely developed, anisotropic physical properties which arise from molecular orientation in the fibre; the bending and relative movements of fibres within yarns, and of yarns themselves intermeshed in three-dimensional structures; the infinite variety of fabric structures, patterns, and textures which can be assembled.

Which really brings us full circle, and why fibres have held the

attention of human beings over past millenia, and will continue to do
so in the future.

Index

'A-Tell', 92
Acetate Rayon, see Rayon, 65
Acrylic fibres, 94 et seq.
  intermediates, 95
  polymerisation, 95
  properties and uses, 96, 116
  structure, 96
Acrylonitrile, 95
Adipic acid, 69
Alginate fibre, 100
Antioxidants, 72, 106
Antistatic agents, 72

Benzene-1,4-dicarboxylic acid, 86
Bicomponent fibres, 47, 120, 123
Blending of fibres, 4, 121
Block polymers, 109
Bonding of fabrics, 123 et seq.

Caprolactam, 73
Carbon fibre, 116
Cellulose, 59
Cellulosic fibres, see Cotton and Rayon, 51, 59 et seq.
  early fibres, 6
Ceramic fibres, 115
Chloroethene (vinyl chloride), 97, 98
Coagulation, 46, 61
Conducting fibres, 78
Copolymers, 9, 94, 97, 98, 109
Cotton
  structure, 25
  preparation of, 49, 122
Crease resistance, 27
Crosslinking, 108
Crystallisation, polymers, 20 et seq., 81
  extended chain, 21, 81
  folded chain, 21
Cyanoethene, 95

Decitex, 16
Delustrant, 34, 72
Denier, 16

Diaminohexane, 69
Dimethylbenzene-1,4-dicarboxylate, 87
Drawing, 40
Dry spinning, 42

Elastomeric fibres, 28, 108 et seq.
Ethanediol, 86
Experimental work, 5, 30

Fabrics, 2, 122
False twist, 119
Fibre B, 81
Fibres
  acrylic, 94 et seq., 116
  bicomponent, 47, 120, 123
  bulked or textured, 2, 118
  carbon, 116
  cellulosic, 51, 59
  ceramic, 115
  chloroethene, 98
  conducting, 78
  crystallisation, 38
  definition, 1
  drawing, 24, 40
  elastic, 108
  glass, 113
  high temperature 80, 93, 113
  history, 6
  hollow, 36, 82
  inorganic, 113
  interaction with water, 17, 26
  metal, 115
  modacrylic, 97 et seq.
  names of, 14
  natural, 25 et seq.
  new, 14, 27, 79, 128
  non-circular, 36
  nylon, see Nylon, 68
  orientation, 24, 26, 81
  polyalkene, 102 et seq.
  polyamide, 68, 79
  polyester, 85 et seq.
  properties, general, 14, 15, 25, 40

134 INDEX

Fibres *continued*
quality, 15, 31, 37, 45, 47
regenerated, 6
relaxation, 26
shrinkage, 26, 56
spandex, 112
spinning technology, 30 et seq.
staple, 2, 121 et seq.
stretching, 24, 45, 47, 62
structure, 18 et seq.
synthetic, 7, 27
vinyl alcohol, 99
vinyl chloride, 98
world production of, 3
Filament, continuous,
definition, 2
spinning, 2, 30 et seq
yarns, 2, 118
Filter pack, 34, 44
Flax, 51

Glass fibres, 113
Glass transition temperature, 24, 26
Grid melter, 32

Hexamethylene diamine, 69
Hexanolactam, 73
Hollow fibres, 36, 82

Inorganic fibres, 113 et seq.

Jute, 51

'Kodel', 91
'Kynol', 92

Linen, 51
'Lycra', 110

Man-made fibre industry
growth of 3, 128
Melded fabrics, 123
Melting of polymers, 24, 28
Melt spinning, 30
Metal fibres, 115
Modacrylic fibres, 97 et seq.
Modifiers, 62
Modulus, 16
Moisture regain, 18

'Nomex', 80
Non-circular fibres, 36, 65
Non-woven fabrics, 106, 122
Nylon, *see also* Polyamides

Nylon *continued*
crystallisation, 38
dye variants, 72
glass transition, 27
intermediates, 69
invention, 8, 68
melt spinning, 38
polymerisation, 71, 74, 76
continuous, 73, 76
properties of, 40, 77
salt, 71
thermal stability, 31
uses, 77
Nylon 4, 82
Nylon 6, 73
Nylon 6.6, 69
Nylon 11, 79
Nylon 12, 79

Orientation of fibres, 24

Plastics, fibre reinforced, 114
Pleat retention, 27
Poly(acrylonitrile), 95
Polyalkene fibres, 102 et seq.
Polyamides, 68 et seq.
chemistry, 8, 69, 72
definitions, 68
for high temperature use, 80
Poly(cyanoethene), 95
Polyesters, 85 et seq., 91
Polyethene, 102
Poly(ethenol) fibre, 99
Poly(ethylene terephthalate)
glass transition, 27
intermediates, 86
invention, 85
melt spinning, 90
polymerisation, 88
properties and uses, 27, 90
thermal stability, 31, 89
Polymerisation
anionic, 12, 104
chain growth, 12
free radical, 10
reactions, 8
step growth, 13
Polymers
and fibres, 14
block, 109
crystallisation, 20, 81
flexibility, 21, 27, 81, 109
isotactic, 12, 103
melting, 24, 27, 31

Polymers *continued*
orientation, 24
relative molecular mass, 10, 13, 20
structure, 8, 11, 19
structural regularity, 11, 19, 109
thermal stability, 31
Polynosic fibres, 64
Polypropene
fibre properties and uses, 106
polymerisation, 103
structure, 103
Polyurethanes, 9, 112
Poly(vinyl alcohol), 99
Poly(vinyl chloride), 97
Propene, 103
Pumps, meter, 34, 44
Pyrrolidone, 82

Rayon, 59 et seq.
acetate, 65
history, 7
secondary acetate, 66
triacetate, 66
viscose, 60
spinning, 61
staple, 64
properties, 64
Reverse osmosis, 82
Rubbers, *see* Elastomeric fibres, 108

Screw extruder, 37
Silk
silkworm, 6, 57
chemistry, 57
spider, 6
Slit film, 106
Spandex fibres, 112
Spherulites, 22
Spinnerets, 34, 47
Spinning
finish, 38
machines, 30 et seq., 40
speeds, 37, 44, 47, 64
technology, 30 et seq.

Spinning technology *continued*
dry, 42
melt, 30
wet, 46, 61
Split film, 106
Staple fibres
definition, 2
non-woven fabrics, 125
spinning, 40, 47, 64
spinning to yarn, 121
Static charging, 78
Stitch bonded fabrics, 127
Stretch yarns, 119

Tenacity, 16
'Terylene', *see* Poly(ethylene terephthalate), 85
Tex, 16, 37
Tyre cords, 17, 62, 81, 113, 116

Vinyl chloride, 97
Viscose rayon, *see* Rayon, 60

Wet spinning, 46, 61
Wool
chemistry of, 53
crimp, 2, 53
felting, 56, 125
preparation of, 55
spinning, 56, 122
structure, 25, 52

Yarns
bulked, textured, 2, 118
continuous filament, 2, 118
definition, 2
general properties, 15
intermingled, 118
spinning, 2
staple, 2, 121
stretch, 119

Ziegler–Natta catalysts, 12, 102, 103